FMEA Using Uncertainty Theories and MCDM Methods

Hu-Chen Liu

FMEA Using Uncertainty Theories and MCDM Methods

Springer

Hu-Chen Liu
School of Management
Shanghai University
Shanghai
People's Republic of China

and

School of Economics and Management
Tongji University
Shanghai
People's Republic of China

ISBN 978-981-10-1465-9 ISBN 978-981-10-1466-6 (eBook)
DOI 10.1007/978-981-10-1466-6

Library of Congress Control Number: 2016939575

Printed on acid-free paper

This Springer imprint is published by Springer Nature
The registered company is Springer Science+Business Media Singapore Pte Ltd.

Preface

Since its introduction by the NASA in 1960s, failure mode and effect analysis (FMEA) has been extensively used to help assure the safety and reliability of products in various industries. Central to FMEA is the prioritization of failure modes based on risk priority number (RPN), which is calculated by the product of the risk factors occurrence (*O*), severity (*S*), and detection (*D*) scaled by experts with an integer number from 1 to 10. However, the conventional RPN method has been criticized as having many inherent deficiencies, thus affecting its effectiveness and limiting its actual applications. In the processes of risk assessment, FMEA team members may not possess a sufficient level of knowledge regarding the risk analysis problem due to the increasing complexity of products, designs, processes, and/or services. In such cases, they usually have some uncertainty in providing their judgments on the identified failure modes, which makes the results of risk evaluation exhibit the characteristics of uncertainty, fuzziness, and imprecision. Besides, the mathematical formula (i.e., multiplication) adopted for determining the failure priority is questionable and lacks adequate scientific basis. For example, the relative weights of risk factors are not taken into account; different combination of *O*, *S*, and *D* ratings may produce the same value of RPN, but their risk implications may be different; the risk factors are evaluated according to discrete ordinal scales of measure, but the calculation of multiplication is meaningless on ordinal scales.

Over the past decades, the improvement of FMEA has been receiving more and more attention from researchers, and a lot of alternative risk priority models have been suggested in the literature to resolve the shortcomings and enhance the performance of the traditional FMEA. First, many uncertainty theories, such as fuzzy set, Dempster–Shafer (D–S) theory, and intuitionistic fuzzy set (IFS), have been utilized to deal with the vagueness and uncertainty in making the criticality assessment. On the other hand, multi-criteria decision-making (MCDM) methods are one of the most popular approaches employed to prioritize the failure modes recognized in FMEA, which can enhance the efficacy and empirical validity of risk assessment results. We remark, despite the existence of other types of FMEA models (such as mathematical programming and artificial intelligence), that the

MCDM-based FMEA under uncertain environment has a series of unique advantages.

The FMEA theory is undergoing continuous in-depth study as well as continuous expansion of the scope of its applications. As such, it has been found that effective assessment and ranking of the failure modes that have been individuated in FMEA becomes increasingly important. Evaluation information modeling and decision-making tools, including uncertainty theories for modeling the ambiguities of risk assessments and MCDM techniques for the priority ranking of failure modes, have broad prospects to improve the criticality analysis process of FMEA, but pose many interesting yet challenging topics for research.

In this book, we will offer a thorough and systematic introduction to the modified FMEA models based on uncertainty theories (e.g., fuzzy logic, IFS, *D* numbers and 2-tuple linguistic variables) and various MCDM methods such as distance-based MCDM, compromise ranking MCDM, hybrid MCDM, etc. The book is structured as the following five parts, which contain 13 chapters.

Part I consists of two chapters (Chaps. 1 and 2), which introduce the traditional FMEA and review the risk evaluation approaches based on uncertainty theories and MCDM methods in FMEA literature. Concretely speaking, Chap. 1 introduces the basics of FMEA, covering its development, implementing procedure, and basic terminology, and summarizes the major shortcomings of the conventional RPN method when applied in practical situations. Chapter 2 makes a comprehensive review of the academic works employing uncertainty theories and MCDM methods to overcome the deficiencies of the traditional FMEA, based on which the current research trends and future research directions in this field of study are also highlighted.

Part II consists of four chapters (Chaps. 3–6), which introduce the FMEA models by using distance-based MCDM methods. Specifically, Chap. 3 introduces the risk assessment methodology for FMEA using intuitionistic fuzzy hybrid weighted Euclidean distance (IFHWED) operator, and illustrates it with an example of developing new horizontal directional drilling (HDD) machine. Chapter 4 introduces the risk priority model for FMEA using interval 2-tuple hybrid weighted distance (ITHWD) measure, and gives its illustration with a case study of blood transfusion. Chapter 5 presents the risk priority model for FMEA based on fuzzy evidential reasoning (FER) and grey relation analysis (GRA) method, and illustrates it by a numerical example. Chapter 6 presents an improved FMEA using *D* Numbers and grey relational projection (GRP) method, and applys it to a case of rotor blades for an aircraft turbine.

Part III consists of two chapters (Chaps. 7–8), which introduce the FMEA models based on compromise ranking MCDM methods. Concretely speaking, Chap. 7 introduces the risk ranking method for FMEA problems, in which fuzzy linguistic terms are used to assess the ratings and weights for risk factors and an extended VIKOR method is used to determine the risk priorities of failure modes. Also, this method is demonstrated with a numerical example concerning the risk analysis in general anesthesia process. Chapter 8 introduces the intuitionistic fuzzy hybrid TOPSIS (IFH-TOPSIS) approach to determine the risk priorities of the failure modes identified in FMEA, and gives a product example of the color super-twisted nematic to show its feasibility and effectiveness.

Part IV consists of three chapters (Chaps. 9–11), which introduce the FMEA frameworks based on other MCDM methods. In Chap. 9, we introduce the risk assessment methodology based on fuzzy decision-making trial and evaluation laboratory (DEMATEL) for the prioritization of failures in system FMEA, and show its application in the thin-film transistor liquid crystal display (TFT-LCD) product. Chapter 10 introduces the risk priority method for FMEA, which uses fuzzy digraph and matrix approach for the risk evaluation of failure modes, and verifies its practicality via a case study of steam valve system. In Chap. 11, we present the FMEA model by applying fuzzy set theory and MULTIMOORA method for failure modes assessment and ranking, and apply it for the prevention of infant abduction in a healthcare facility.

Part V consists of two chapters (Chaps. 12–13), which introduce the FMEA approaches by utilizing hybrid MCDM methods. Specifically, Chap. 12 introduces the hybrid MCDM method for risk analysis based on combination weighting and fuzzy VIKOR method, in which fuzzy analytic hierarchy process (AHP) is combined with entropy method for risk factor weighting. Furthermore, this FMEA method is applied for analyzing the risk of general anesthesia process to illustrate its feasibility and applicability. Chapter 13 introduces the hybrid MCDM method for FMEA that uses a modified VIKOR to determine the effects of failure modes, the DEMATEL to construct the influential relations among failure modes and causes of failures, and the AHP approach to obtain the prioritization levels for failure modes. Finally, a numerical example concerning diesel engine turbocharger system is given to demonstrate the FMEA approach being proposed.

This book is useful for practitioners and researchers working in the fields of quality management, decision making, information science, and management science and engineering. It can also be used as a textbook for postgraduate and senior undergraduate students.

This work was supported by the National Natural Science Foundation of China (No. 71402090), the NSFC key program (No. 71432007), the China Postdoctoral Science Foundation (No. 2015T80456), and the Program for Young of Special Appointment (Eastern Scholar) at Shanghai Institutions of Higher Learning (No. QD2015019). Finally, I am grateful to my family for their constant love, encouragement, and support.

Shanghai
March 2016

Hu-Chen Liu

Contents

Part I
FMEA and Its Improvements

Chapter 1
FMEA

Failure mode and effect analysis (FMEA), first developed as a formal design methodology in the 1960s by the aerospace industry, is a systematic methodology designed to identify known and potential failure modes and their causes, and the effects of failure on the system or end users, to assess the risk associated with the identified failure modes and prioritize them for proactive interventions, and to carry out corrective actions for the most serious issues to enhance the reliability and safety of products and processes, designs, or services. Traditionally, criticality or risk assessment in FMEA is carried out via the risk priority number (RPN), made up of the arithmetic product of occurrence (*O*), severity (*S*), and detection (*D*). FMEA has been proven to be a useful and powerful tool in assessing potential failures and preventing them from occurring. In this chapter, we first introduce the basics of FMEA, including its fundamental concepts, development, implementing procedure, basic terminology. Besides, the important weaknesses of the crisp RPN method when applied in the real-world cases are presented finally.

1.1 The Traditional FMEA

FMEA is an engineering technique used to define, identify, and eliminate known and/or potential failures, problems, errors, and so on from the system, design, process, and/or service before they reach the customer (Stamatis 2003). When it is used for a criticality analysis, it is also referred to as failure mode, effect, and criticality analysis (FMECA). The main objective of FMEA is to identify potential failure modes, evaluate the causes and effects of different component failure modes, and determine what could eliminate or reduce the chance of high-risk failures (Liu et al. 2015a). The results of the analysis can help risk analysts to identify and correct the failure modes that have a detrimental effect on the system and improve its performance during the stages of design and development.

H.-C. Liu, *FMEA Using Uncertainty Theories and MCDM Methods*,
DOI 10.1007/978-981-10-1466-6_1

FMEA has been around for many decades as a powerful tool to support product designs, manufacturing processes, services, and maintenances. It was originally created and developed in the USA in the early 1960s by the aerospace industry during the Apollo mission to evaluate the impact of system and equipment failures on mission success, personnel and system safety, maintainability, and system performance (Liu et al. 2015b). In the late 1970s, the Ford Motor Company introduced FMEA to the automotive industry for safety and regulatory consideration and used it to improve production and design. In 1980, the FMEA implementation process was standardized by the Military Standard (MIL-STD) 1629A. In 1990, the International Organization for Standardization (ISO) recommended the use of FMEA for design review in the ISO 9000 series. In 1994, jointly developed by Chrysler, Ford, and General Motors, the first version of the standard Society of Automotive Engineers (SAE) J1739 was published. This document describes the use of FMEA and gives general guidance in the application of design and process FMEAs. Today, FMEA has been extensively used as a powerful tool for safety and reliability analysis of products and processes in a wide range of industries, including aerospace, automotive, nuclear, electronics, chemical, mechanical, and health care, to name a few (Liu et al. 2013, 2015a, 2016a, b).

In order to analyze a specific product or system, a cross-functional team of domain experts should be established for carrying out FMEA first. The next step in FMEA is to identify all possible potential failure modes of the product or system by a session of systematic brainstorming (McDermott et al. 2009). After that, criticality analysis is performed on these failure modes taking into account the risk factors for occurrence (O), severity (S), and detection (D). The main purpose of FMEA is to allow the analysts to prioritize the failure modes of a system, design, process, product, or service in order to assign the limited resources to the highest risk items.

Traditionally, the prioritization of failure modes in FMEA is determined through the risk priority number (RPN), which is defined as the multiplication of the risk factors O, S, and D for each failure. That is,

$$\mathrm{RPN} = O \times S \times D, \tag{1.1}$$

where O is the probability or frequency of the failure, S is the seriousness (consequence) of the failure, and D is the ability to detect the failure before the impact of the effect is realized. For obtaining the RPN of a potential failure mode, each of the three risk factors is usually rated on a numerical scale ranging from 1 to 10. An example ranking system for the three risk factors is provided in Tables 1.1, 1.2 and 1.3 (Liu et al. 2012, 2013). The higher the RPN of a failure mode, the greater the risk is for product/system reliability. With respect to the scores of RPNs, the failure modes can be ranked, and then, proper actions will be preferentially taken on the high-risk failure modes. RPNs should be recalculated after the corrections to see whether the risks have gone down and to check the efficiency of the corrective precaution for each failure mode. FMEA is a dynamic document, changing as the system, design, process, product, and/or service changes with the intent always to make a better system, design, process, product, and/or service (Stamatis 2003).

Table 1.1 Traditional ratings for occurrence of a failure mode

Rating	Probability of failure	Possible failure rate
10	Extremely high: failure almost inevitable	≥1 in 2
9	Very high	1 in 3
8	Repeated failures	1 in 8
7	High	1 in 20
6	Moderately high	1 in 80
5	Moderate	1 in 400
4	Relatively low	1 in 2000
3	Low	1 in 15,000
2	Remote	1 in 150,000
1	Nearly impossible	≤1 in 1,500,000

Table 1.2 Traditional ratings for severity of a failure mode

Rating	Effect	Severity of effect
10	Hazardous without warning	Highest severity ranking of a failure mode, occurring without warning, and consequence is hazardous
9	Hazardous with warning	Higher severity ranking of a failure mode, occurring with warning, and consequence is hazardous
8	Very high	Operation of system or product is broken down without compromising safe
7	High	Operation of system or product may be continued, but performance of system or product is affected
6	Moderate	Operation of system or product is continued, and performance of system or product is degraded
5	Low	Performance of system or product is affected seriously, and the maintenance is needed
4	Very low	Performance of system or product is less affected, and the maintenance may not be needed
3	Minor	System performance and satisfaction with minor effect
2	Very minor	System performance and satisfaction with slight effect
1	None	No effect

1.2 The Procedure of FMEA

To carry out an FMEA effectively, a systematic approach should be followed. The general procedure for conducting an FMEA can be divided into several steps as shown in Fig. 1.1. These steps are briefly explained here (Pillay and Wang 2003; Liu et al. 2011, 2014):

Step 1. Determine the scope of FMEA analysis

By definition, the FMEA is a specific methodology to evaluate a system, design, process, or service for possible ways in which failures can occur. Thus, the first step is to define the particular scope of an individual FMEA for narrowing down the

Table 1.3 Traditional ratings for detection of a failure mode

Rating	Detection	Criteria
10	Absolutely impossible	Design control does not detect a potential cause of failure or subsequent failure mode, or there is no design control
9	Very remote	Very remote chance the design control will detect a potential cause of failure or subsequent failure mode
8	Remote	Remote chance the design control will detect a potential cause of failure or subsequent failure mode
7	Very low	Very low chance the design control will detect a potential cause of failure or subsequent failure mode
6	Low	Low chance the design control will detect a potential cause of failure or subsequent failure mode
5	Moderate	Moderate chance the design control will detect a potential cause of failure or subsequent failure mode
4	Moderately high	Moderately high chance the design control will detect a potential cause of failure or subsequent failure mode
3	High	High chance the design control will detect a potential cause of failure or subsequent failure mode
2	Very high	Very high chance the design control will detect a potential cause of failure or subsequent failure mode
1	Almost certain	Design control will almost certainly detect a potential cause of failure or subsequent failure mode

project focus. Giving clear and careful thought to this step is very vital because clearly defined boundaries establish the issues that are to be considered and the approaches that the team will take during the analysis.

Step 2. Assemble the FMEA team

FMEA is a team-based activity and cannot be done by one person alone. So considering the FMEA problem scope defined in previous step, we must form a correct team of subject matter experts from a variety of disciplines. All team members must have some knowledge of group behavior, the task at hand, the problem to be discussed, and direct or indirect ownership of the problem. Someone possessing, or well trained in, team facilitation skills should lead the FMEA team. It is important to note that the team must be cross-functional and multi-disciplined.

Step 3. Understand the system to be analyzed

One of the most important steps in FMEA is to understand the system to be analyzed. It needs to divide the system into subsystems and/or assemblies and use blue prints, schematics, and flowcharts to identify components and relations among components. For system and design FMEAs, the functional block diagram, parameter diagram (P diagram), and FMEA interface matrix are applicable. For the process and service FMEAs, the process flowchart and process flow diagram worksheet are applicable.

Step 4. Brainstorm failure modes of each component and their effects

Once everyone on the FMEA team has an understanding of the system, a series of brainstorming sessions should be conducted to list all the potential failure modes

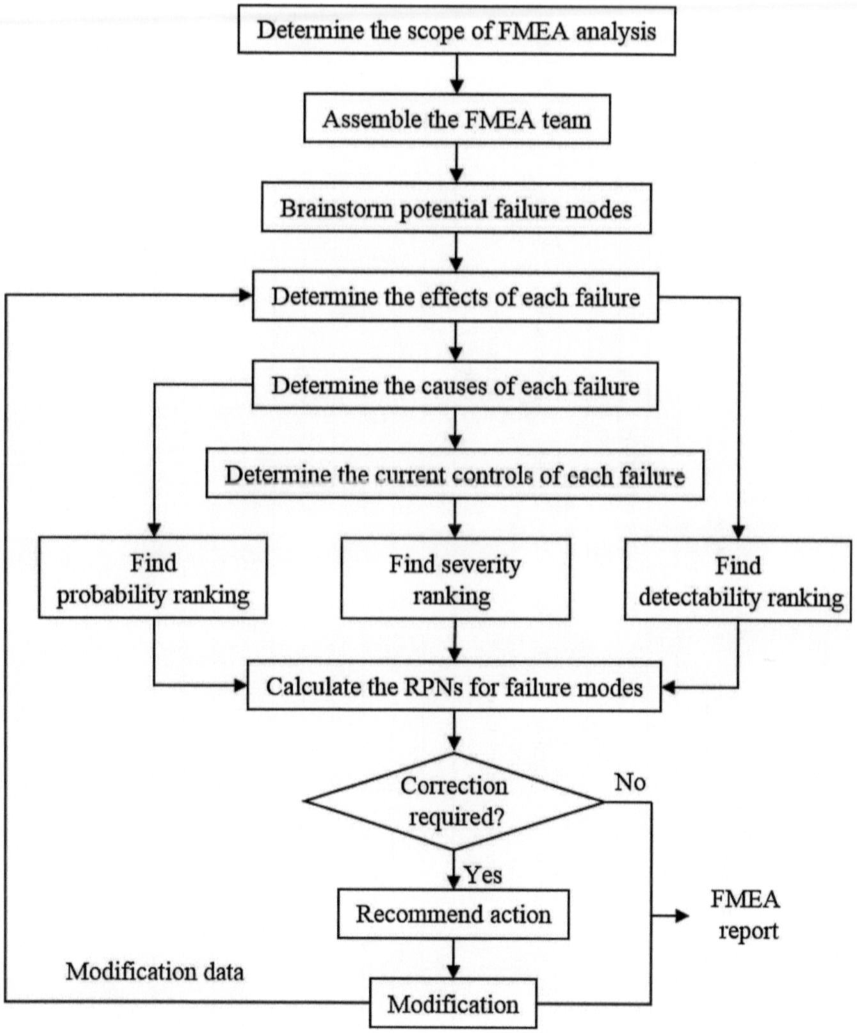

Fig. 1.1 Main steps of FMEA

that could affect the product quality and identify the potential effects of the failure should it occur. Note that there are usually several failure modes for each component. For some of the failure modes, there may be only one effect, while for other failures, there may be several effects.

Step 5. Determine the *O*, *S*, and *D* for failure modes

Normally, the three risk factors *O*, *S*, and *D* are ranked based on a 10-point scale, with 1 being the lowest and 10 the highest. Establishing clear and concise descriptions for the points on each of the three rating scales is important, so that all team members have the same understanding of the rankings. Also, it is important to tailor the risk ranking scales to organization-specific applications.

Table 1.4 Format of an FMEA worksheet

FMEA report																
	System:								FMEA No.:							
	Subsystem:								Page:							
	Component:								FMEA date (original):							
	FMEA team:								(revised):							
	FMEA process								Action results							
Item	Component/process	Potential failure mode	Potential effect(s) of failure	*S*	Potential cause(s) of failure	*O*	Current control mechanisms	*D*	RPN	Recommended action(s)	Responsibility and completion date	Action taken	*S*	*O*	*D*	RPN

Step 6. Calculate the RPN of each failure mode

The RPN is simply calculated by multiplying the risk factors representing the probability, severity, and detectability for each item. The total RPN of FMEA can also be calculated by adding the RPNs of all individuated failure modes in order to compare the updated total RPN value once the recommended actions have been instituted.

Step 7. Prioritize the failure modes for preventive actions

The failure modes can now be prioritized by ranking them in terms of the RPNs in decreasing order. Then, recommended actions for the high-risk failure modes should be developed to enhance the system performance. Normally, these actions fall into three categories: eliminating failure modes, increasing failure detectability, and minimizing losses in the event that a failure occurs.

Step 8. Prepare FMEA report by summarizing the analysis results

The FMEA process should be documented using an FMEA worksheet as shown in Table 1.4. This form captures all of the important information regarding FMEA and serves as an excellent communication tool.

Step 9. Calculate the revised RPNs as the failure modes are reduced or eliminated

Once the recommended actions have been taken to improve the system, the FMEA team must reassess each of the risk rankings for O, S, and D and pursue improvement all over again. This risk reassessment is very important because it shows how well the risk associated with each failure mode is reduced as a result of the specific actions from the FMEA. The long-term goal is to completely eliminate every single failure, and the short-term goal is to minimize the failures if not eliminate them.

1.3 The Terminology in FMEA

Although there have been many variations of FMEA, the specific words used and their special meaning throughout the years have been maintained. Some of the terms commonly used in conducting an FMEA are introduced below. For a more detailed vocabulary list of FMEA, please see Stamatis (2003), and Carlson (2012).

Function. It is the task that the system, design, process, component, subsystem, and service must perform. For design FMEA, this is the primary purpose or design intent of the item. For process FMEA, this is the primary purpose of the manufacturing or assembly operation along with any needed requirement. There may be many functions for each item or operation.

Failure mode. Failure mode is the physical description of the manner in which a failure occurs. A failure describes the way in which a product could fail to perform its desired intent as described by the needs, wants, and expectations of the customers. A failure mode may have more than one level depending on the complexity of the defined system, design, process, or service.

Effect of failure. An effect is an adverse consequence of the failure on the system, design, process, or service. It should be noted that the effect of a failure can be addressed from two points of view. The first one is local, in which the failure is isolated and does not affect anything else. The second one is global, in which the failure can and does affect other functions and/or components. It has a domino effect. In general, the failure with a global effect is more serious than the one with a local nature.

Causes of failure. This is a list conceivable root causes assignable to each failure mode. The causes listed should be concise and as complete as possible. For design FMEA, the cause is the design deficiency that results in the failure mode. For process FMEA, the cause is the manufacturing or assembly deficiency that results in the failure mode. In most applications, particularly at the component level, the cause is taken to the level of failure mechanism. There can be many causes for each failure mode.

Current controls. They are methods or actions currently planned or that are already in place, to reduce or eliminate the risk associated with each potential cause of failure. Controls can be the methods to prevent or detect the cause during product development, or actions to detect a problem during service before it becomes catastrophic.

Recommended actions. They are specific actions recommended by the FMEA team that can be implemented to reduce or eliminate the risk associated with potential cause(s) of each failure mode. Recommended actions should consider the existing controls, the relative importance (prioritization) of the issues, and the cost and effectiveness of the corrective actions.

1.4 Shortcomings of the Traditional FMEA

The traditional FMEA has been proven to be a useful risk assessment tool and one of the most important early preventative management initiatives that help examine known or potential failures to prevent them from happening. However, the conventional RPN calculation method has been considerably criticized in the literature for a vast variety of reasons. The most important shortcomings reported in the FMEA literature are summarized as follows (Liu et al. 2013):

(1) The relative importance among O, S, and D is not taken into account. The three risk factors are treated with the same weight. This may not be the case when considering a practical application of FMEA.

(2) Different combination of O, S, and D may produce exactly the same value of RPN, but their hidden risk implications may be totally different. This could entail a waste of resources and time, or in some cases, some high-risk failure modes gone unnoticed.

(3) The three risk factors are difficult for FMEA team members to precisely determine. Much information in FMEA is often uncertain or vague and can

be expressed in a linguistic way such as "likely," "important," or "very high."

(4) The mathematical formula for calculating RPN is debatable and lacks a complete scientific basis. There is no rationale as to why *O*, *S*, and *D* should be multiplied to produce the RPN.

(5) The direct and indirect relationships between failure modes and causes of failure are not taken into consideration.

(6) The three risk factors *O*, *S*, and *D* are evaluated according to discrete ordinal scales of measure. But that the calculation of multiplication is meaningless on ordinal scales.

(7) The RPN considers only three risk factors, mainly in terms of safety. Other important risk factors such as economical aspects are ignored.

(8) The conversion of scores is different for the three risk factors. The relationship between *O* and the associated ratings is nonlinear, while the *D* and the associated ratings have a linear relationship.

(9) The RPN values are not continuous with many holes. Many of the numbers in the range of 1–1, 000 cannot be formed from the product of *O*, *S*, and *D*, and only 120 of the 1000 numbers can be generated from the multiplication of risk factors.

(10) The mathematical form adopted for calculating RPN is strongly sensitive to the variation in risk factor evaluations. Small variation in one rating may lead to vastly different effects on the RPN value.

(11) The RPN method is only measuring from the risk viewpoints while ignoring the importance of corrective actions. It cannot be used to measure the effectiveness of corrective actions.

(12) The RPN scale itself has some non-intuitive statistical properties. The initial and correctly assumed observation that the scale starts at 1 and ends at 1000 often leads to incorrect assumptions regarding the middle of the scale.

For comprehensive reviews on the drawbacks of the traditional FMEA, one can refer to Appendix.

References

Carlson C (2012) Effective FMEAs: achieving safe, reliable, and economical products and processes using failure mode and effects analysis. Wiley, New Jersey

Liu HC, Liu L, Bian QH, Lin QL, Dong N, Xu PC (2011) Failure mode and effects analysis using fuzzy evidential reasoning approach and grey theory. Expert Syst Appl 38(4):4403–4415

Liu HC, Liu L, Liu N, Mao LX (2012) Risk evaluation in failure mode and effects analysis with extended VIKOR method under fuzzy environment. Expert Syst Appl 39(17):12926–12934

Liu HC, Liu L, Liu N (2013) Risk evaluation approaches in failure mode and effects analysis: a literature review. Expert Syst Appl 40(2):828–838

Liu HC, Fan XJ, Li P, Chen YZ (2014) Evaluating the risk of failure modes with extended MULTIMOORA method under fuzzy environment. Eng Appl Artif Intell 34:168–177

Liu HC, You JX, Ding XF, Su Q (2015a) Improving risk evaluation in FMEA with a hybrid multiple criteria decision making method. Int J Qual Reliab Manage 32(7):763–782

Liu HC, You JX, Shan MM, Shao LN (2015b) Failure mode and effects analysis using intuitionistic fuzzy hybrid TOPSIS approach. Soft Comput 19(4):1085–1098

Liu HC, You JX, Chen SM, Chen YZ (2016a) An integrated failure mode and effect analysis approach for accurate risk assessment under uncertainty. IIE Trans. doi:10.1080/0740817X.2016.1172742

Liu HC, You JX, Li P, Su Q (2016b) Failure mode and effect analysis under uncertainty: an integrated multiple criteria decision making approach. IEEE Trans Reliab (In press)

McDermott RE, Mikulak RJ, Beauregard M (2009) The basics of FMEA. 2nd edn. Taylor & Francis, New York

Pillay A, Wang J (2003) Modified failure mode and effects analysis using approximate reasoning. Reliab Eng Syst Saf 79(1):69–85

Stamatis DH (2003) Failure mode and effect analysis: FMEA from theory to execution. 2nd edn. ASQC Press, New York

Chapter 2
FMEA Using Uncertainty Theories and MCDM Methods

To resolve the shortcomings of the conventional RPN method, a great number of studies have been conducted on the improvement of FMEA and a variety of alternative approaches have been proposed. Liu et al. (2013) first reviewed the approaches employed to enhance the performance of FMEA and proposed a framework to categorize the improved models for risk evaluation in FMEA. In this chapter, we update the results in Liu et al. (2013) and provide a systematic review of those academic works attempting to deal with the deficiencies of the traditional FMEA based on uncertainty theories and MCDM methods. Related articles appearing in the international journals from 1992 to 2016 are gathered and analyzed. Based on the 64 journal articles collected, the specific objectives of this chapter are to summarize the MCDM methods that have been used in the FMEA literature and to show the current research trends and future research directions in this field of study.

2.1 Research Methodology

This chapter presents the results of an extensive literature survey on the risk evaluation methods in FMEA using uncertainty theories and MCDM methods. For this purpose, we searched in Scopus database for academic articles published between 1992 and 2016. The keywords "FMEA" and "failure mode and effect analysis" were used for searching in "abstract, title, and keywords" for journal papers. As a result, a total of 849 document results were identified from the Scopus database, of which 64 fall under the scope of this review after the title, abstract, and full-text screening. Publications in languages other than English and non-refereed professional publications, such as textbooks, doctoral dissertations, and conference proceedings, were not included. Furthermore, we only included articles that

H.-C. Liu, *FMEA Using Uncertainty Theories and MCDM Methods*,
DOI 10.1007/978-981-10-1466-6_2

reported on a MCDM method or technique that specifically aims at overcoming some of the shortcomings of the traditional FMEA.

A variety of MCDM-based risk priority models are found in the literature to improve the criticality analysis process of FMEA. Depending on the MCDM methods used, the reviewed papers are roughly grouped into four categories: the ones using distance-based MCDM methods (16 articles), the ones using compromise ranking MCDM methods (11 articles), the ones using outranking MCDM methods (3 articles), the ones using pairwise comparison MCDM methods (8 articles), the ones using other MCDM methods (18 articles), and the ones based on hybrid MCDM methods (8 articles). The classification scheme of these FMEA articles is shown in Table 2.1. In what follows, we more specifically go into the references and show what has been done.

Table 2.1 Classification of MCDM-based FMEA models

Categories	Approaches	Reference	Percentage (%)
Distance-based MCDM methods	Distance measure	Liu et al. (2014c, e)	25.0
	GRA	Geum et al. (2011), Liu et al. (2015a, d), Chang et al. (1999, 2001), Pillay and Wang (2003), Sharma et al. (2007, 2008), Sharma and Sharma (2012, 2015), Moon et al. (2013), Tsai and Yeh (2015); Panchal and Kumar (2016), Zhou and Thai (2016)	
Compromise ranking MCDM methods	VIKOR	Liu et al. (2012), Safari et al. (2016), Emovon et al. (2015)	17.2
	TOPSIS	Braglia et al. (2003), Helvacioglu and Ozen (2014), Sachdeva et al. (2009), Song et al. (2013, 2014), Chang (2015), Liu et al. (2015d), Vahdani et al. (2015)	
Outranking MCDM methods	QUALIFLEX	Liu et al. (2016b)	4.7
	ELECTRE	Liu et al. (2016a)	
	PROMETHEE	Lolli et al. (2015)	
Pairwise comparison MCDM methods	AHP	Braglia (2000), Chang (2016), Carmignani (2009), Hu et al. (2009), Ilangkumaran et al. (2014), Abdelgawad and Fayek (2010)	12.5
	ANP	Zammori and Gabbrielli (2011), Hsu et al. (2013)	

(continued)

Table 2.1 (continued)

Categories	Approaches	Reference	Percentage (%)
Other MCDM methods	ME-MCDM	Franceschini and Galetto (2001)	28.1
	GFEA	Jenab and Dhillon (2005)	
	MULTIMOORA	Liu et al. (2014b), Zhao et al. (2016)	
	COPRAS-G	Adhikary et al. (2014)	
	ER	Chin et al. (2009), Gargama and Chaturvedi (2011), Du et al. (2016)	
	DEMATEL	Seyed-Hosseini et al. (2006), Chang (2009), Chang and Cheng (2010), Chang and Cheng (2011), Chang (2014), Li et al. (2012, 2015c), Wang et al. (2016)	
	Digraph and matrix	Gandhi and Agrawal (1992), Liu et al. (2014a)	
Hybrid MCDM methods	AHP and TOPSIS	Kutlu and Ekmekçioğlu (2012), Ekmekçioğlu and Kutlu (2012)	12.5
	VIKOR and AHP	Liu et al. (2015e)	
	VIKOR, DEMATEL, and AHP	Liu et al. (2015b)	
	FER and GRA	Liu et al. (2011)	
	ER and TOPSIS	Du et al. (2014)	
	GRA and DEMATEL	Chang et al. (2013)	
	TOPSIS and DEMATEL	Chang et al. (2014)	

2.2 MCDM Methods for FMEA

2.2.1 Distance-Based MCDM Methods

Liu et al. (2014c) developed an alternative risk assessment methodology using the intuitionistic fuzzy hybrid weighted Euclidean distance (IFHWED) operator for risk evaluation and prioritization of failure modes in FMEA. Particularly, both subjective and objective weights of risk factors are considered using the developed method. Liu et al. (2014e) evaluated the risk of failure modes by the interval 2-tuple hybrid weighted distance (ITHWD) measure and proposed a new risk priority to improve the performance of the traditional FMEA. The new model can not only consider the subjective and objective weights of risk factors but also handle the uncertainty and diversity of FMEA team members' assessment information in the risk prioritization process.

Geum et al. (2011) proposed a systematic approach for identifying and evaluating potential failures based on service-specific FMEA and grey relational analysis (GRA) method. Firstly, the service-specific FMEA was constructed to incorporate the multilateral service-specific characteristics to FMEA. Next, the GRA was applied for calculating the risk score of each dimension (i.e., *O*, *S*, and *D*) and the final risk priority of each service failure mode. Liu et al. (2015a) presented a novel FMEA approach combining interval 2-tuple linguistic variables with GRA to capture FMEA team members' diversity assessments and improve the effectiveness of the traditional FMEA, and Liu et al. (2014d) proposed a new risk priority model for the risk evaluation in FMEA based on D numbers and an improved GRA method, called grey relational projection (GRP).

Chang et al. (1999) proposed a FMEA approach for finding the RPNs based on fuzzy method and grey theory, where the GRA is applied to determine the risk priority of potential causes. Chang et al. (2001) also utilized the grey theory for FMEA, but the degrees of relational were computed through the traditional crisp scores 1–10 for the risk factors *O*, *S*, and *D*. Other applications of GRA method for the prioritization of failure modes in FMEA can be found in Pillay and Wang (2003), Sharma et al. (2007, 2008), Sharma and Sharma (2012, 2015), Moon et al. (2013), Tsai and Yeh (2015), Panchal and Kumar (2016), Zhou and Thai (2016).

2.2.2 Compromise Ranking MCDM Methods

Liu et al. (2012) determined the risk priorities of failure modes with an extended VIKOR method under fuzzy environment. In this study, the linguistic terms expressed in trapezoidal fuzzy numbers were used to assess the ratings for risk factors and an extension of the VIKOR was utilized to determine risk priorities of the identified failure modes in FMEA. Because of the drawbacks of the traditional FMEA, Safari et al. (2016) employed the fuzzy VIKOR-based FMEA to evaluate enterprise architecture (EA) risks to facilitate EA deployment in an organization, and Emovon et al. (2015) used an enhanced FMEA model integrating an averaging technique with the VIKOR to prioritize the risk of failure modes for marine machinery systems.

Braglia et al. (2003) first used the fuzzy technique for order preference by similarity to ideal solution (TOPSIS) approach for prioritizing failure modes in FMEA, in which the ranking for failure causes is determined based on the measurement of the Euclidean distance of an alternative from an ideal goal. Helvacioglu and Ozen (2014) proposed a risk priority framework to overcome the shortcomings of the traditional FMEA through fuzzy TOPSIS and applied it to yacht system design. Based on the TOPSIS, Sachdeva et al. (2009) presented an alternative FMEA approach for prioritizing failure modes, which considers the risk factors for failure occurrence, non-detection, maintainability, spare parts, economic safety, and economic cost and employs the Shannon entropy concept to compute the objective weights of the six risk factors. Song et al. (2013) developed a failure evaluation structure based on fuzzy TOPSIS and comprehensive weighting method to improve the effectiveness of FMEA technique, and Song et al. (2014) proposed a FMEA approach using rough set theory and group TOPSIS method for ranking the risk of failure modes under subjective and uncertain environment.

Liu et al. (2015d) introduced a new modified TOPSIS method, namely the intuitionistic fuzzy hybrid TOPSIS approach, to determine the risk priorities of the failure modes identified in FMEA. Vahdani et al. (2015) proposed a modified version of FMEA by integrating fuzzy belief structure and TOPSIS to alleviate the drawbacks and improve the risk evaluation process of the traditional FMEA. Chang (2015) proposed a risk assessment method based on soft TOPSIS approach to solve the risk assessment problem in FMEA under a linguistic environment.

2.2.3 *Outranking MCDM Methods*

Liu et al. (2016a) described the application of an ELECTRE-based outranking approach for FMEA within the interval 2-tuple linguistic environment. Considering different types of FMEA team members' assessment information, a hybrid averaging operator was employed to construct the group assessment matrix and a modified ELECTRE method was used to analyze the group interval 2-tuple linguistic data to determine the risk ranking of failure modes. Liu et al. (2016b) developed a new risk priority model for FMEA by integrating hesitant 2-tuple linguistic term sets and an extended QUALIFLEX approach. In this model, the concept of hesitant 2-tuple linguistic term sets was presented to express various uncertainties in the assessment information of FMEA team members and a multiple objective optimization model based on GRA method was constructed to determine the relative weights of risk factors with incomplete weight information. Finally, the extended QUALIFLEX approach with an inclusion comparison method was suggested for prioritizing failure modes incorporating interrelationship between failure modes, cost of failure, and corrective action cost as additional risk factors. Lolli et al. (2015) proposed a MCDM method for FMEA based on the PROMETHEE notation to sort failure modes into priority classes, in which FMEA team members are asked to establish the reference profiles on each risk factor according to their experience and skills for obtaining the global classification of the failure modes.

2.2.4 *Pairwise Comparison MCDM Methods*

Braglia (2000) developed a multi-attribute failure mode analysis (MAFMA) model based on analytic hierarchy process (AHP) technique, which considers the risk factors (O, S, D, and expected cost) as decision criteria, possible causes of failure as decision alternatives, and the priority ranking of failure causes as decision goal. Then, following the AHP procedure, all the possible causes of failure are evaluated and ranked. Making reference to Braglia (2000), Chang (2016) proposed an approach that integrates MAFMA and the 2-tuple representation method for risk assessment and prioritization. Carmignani (2009) presented a priority-cost-based FMEA approach based on a new interpretation of RPN, the AHP technique, and the new variable of profitability, in which the AHP is used to determine the different weights of risk factors. Hu et al. (2009) utilized fuzzy AHP to determine the relative weights of risk factors and proposed a green component risk priority number (GC-RPN) to analyze the risk of green components to hazardous substance, and Ilangkumaran et al. (2014) used fuzzy AHP to compute the risk factor weights and developed an evaluation model based on FMEA and fuzzy AHP for assessing the risk priority of the critical components in a paper industry. Abdelgawad and Fayek (2010) used fuzzy expert system and fuzzy AHP to address the limitations of the traditional calculation of RPN and extended the application of FMEA to risk criticality assessment in the construction industry.

Zammori and Gabbrielli (2011) presented a risk assessment procedure by integrating FMEA and analytic network process (ANP) taking into account possible interactions among the principal causes of failure in the criticality assessment. According to the model, *O*, *S*, and *D* were split into subcriteria and arranged in a hybrid (hierarchy/network) decision structure to compute the RPN. The causes of failure were included in the lowest level of the structure, and their effects and the strengths of their dependencies were assessed via pairwise judgments. Hsu et al. (2013) utilized the FMEA to construct a materiality analysis model for determining material issues in sustainability reporting, in which the ANP is employed to determine the weights of risk factors and a RPN of materiality analysis is calculated for material issues of sustainability reporting to rank them in accordance with stakeholder needs.

2.2.5 Other MCDM Methods

Franceschini and Galetto (2001) presented a multi-expert MCDM (ME-MCDM) method to calculate the risk priority levels of the failure modes in FMEA, which is able to deal with the information provided by the design team without necessitating an arbitrary and artificial numerical conversion. Jenab and Dhillon (2005) reported a group-based failure effect analysis (GFEA) method to mitigate the problems of the conventional RPN approach, which uses group decision-making technique to study the failure risk category with uncertain information and uses the compensated operators to allow the trade-off among risk factors.

Liu et al. (2014b) proposed a risk priority model for evaluating the risk of failure modes based on fuzzy set theory and MULTIMOORA method, in which the risk factors and their relative weights are evaluated using fuzzy ratings, and the failure modes are ranked through an extended MULTIMOORA method. Zhao et al. (2016) presented a new approach for FMEA based on interval-valued intuitionistic fuzzy sets (IVIFSs) and MULTIMOORA method to handle the uncertainty and vagueness from FMEA team members' subjective assessments and to get a more accurate ranking of failure modes identified in FMEA. In the proposed approach, the interval-valued intuitionistic fuzzy continuous weighted entropy was applied for risk factor weighting. Adhikary et al. (2014) substituted the conventional RPN estimation method by grey complex proportional assessment (COPRAS-G) and presented a multi-criteria FMEA for coal-fired thermal power plants using uncertain data. In this model, the weight of each risk factor was calculated based on the Shannon entropy concept.

Chin et al. (2009) presented a risk priority model for FMEA using the group-based evidential reasoning (ER) approach to capture FMEA team members' diversity judgments and prioritize failure modes under different types of uncertainties such as incomplete assessment, ignorance, and intervals. The belief structures in evidence theory were also used by Gargama and Chaturvedi (2011) to handle the diversity and uncertainty in the opinions of FMEA team members. To improve the model of Chin et al. (2009), Du et al. (2016) proposed an evidential downscaling method to make FMEA more efficient in practical applications.

Seyed-Hosseini et al. (2006) used the decision-making trial and evaluation laboratory (DEMATEL) technique for reprioritization of failure modes in a system FMEA. The proposed methodology prioritizing failures in terms of direct/indirect relationships between them is suitable for large systems with many subsystems or components. Later, Chang (2009) proposed a methodology which combines the ordered weighted geometric averaging (OWGA) operator and the DEMATEL to evaluate the risk of failures in FMEA; Chang and Cheng (2010) proposed an algorithm integrating intuitionistic fuzzy sets (IFSs) and the DEMATEL for prioritization of failures, Chang and Cheng (2011) suggested an approach which utilizes fuzzy ordered weighted averaging (OWA) operator and the DEMATEL to evaluate the orderings of risk for failure problems, and Chang (2014) presented a soft set-based DEMATEL technique for the prioritization of failures in a product FMEA. To handle uncertain or vague data sets, Li et al. (2012) proposed a methodology by combining evidence theory, IFS, and the DEMATEL approach to make risk assessment for an FMEA system. Recently, Liu et al. (2015c) proposed an integrated approach for FMEA based on fuzzy weighted average (FWA) and fuzzy DEMATEL that could not only cope with the interdependencies among various failure modes but also avoid the shortcomings of the previous DEMATEL-based risk assessment methods. Wang et al. (2016) determined the risk factor weights with the combined entropy and expert evaluation method, then prioritized failure modes by using the DEMATEL method, and finally designed an improved FMECA for feed system of CNC machining center.

Gandhi and Agrawal (1992) presented a method for FMEA of mechanical and hydraulic systems based on a digraph and matrix approach by considering structural and functional interaction of the system. Liu et al. (2014a) developed a novel FMEA model, which uses fuzzy digraph and matrix approach for risk evaluation and prioritization of failure modes. This model first developed a risk factor fuzzy digraph considering risk factors and their relative importance, then formed corresponding fuzzy risk matrixes for all the failure modes in FMEA, and finally computed the risk priority indexes (RPIs) to determine the risk priority of the failure modes.

2.2.6 *Hybrid MCDM Methods*

Kutlu and Ekmekçioğlu (2012) considered a fuzzy approach for FMEA by applying fuzzy TOPSIS integrated with fuzzy AHP, in which the fuzzy AHP is applied to determine the weight vector of risk factors and the fuzzy TOPSIS is adopted to get the risk ranking orders of failure modes. Ekmekçioğlu and Kutlu (2012) further applied the fuzzy hybrid FMEA approach based on fuzzy TOPSIS and fuzzy AHP to a spindle manufacturing process. Liu et al. (2015e) presented a risk evaluation methodology for FMEA based on combination weighting and fuzzy VIKOR, in which integration of fuzzy AHP and entropy method is utilized for risk factor weighting and fuzzy VIKOR method is used to obtain the risk priorities of the identified failure modes. Liu et al. (2015b) combined the VIKOR, the DEMATEL, and the AHP to develop a hybrid MCDM method for FMEA, which used a modified VIKOR method to determine the effects of failure modes together and the DEMATEL technique in conjunction with AHP to construct the influential relation map among failure modes and determine the prioritization level for the failure modes.

Liu et al. (2011) reported a risk priority model using fuzzy evidential reasoning (FER) approach and grey theory to improve the effectiveness of the traditional FMEA. In this model, the FER approach was employed to capture FMEA team members' diversity opinions under different types of uncertainties, and the GRA method was used to determine the risk priorities of the failure modes that have been identified. In Du et al. (2014), the authors applied ER approach to express FMEA team members' assessment information, employed the TOPSIS to acquire the risk priority of failure modes, and finally provided a fuzzy FMEA method using the ER and the TOPSIS.

Chang et al. (2013) proposed an approach based on the GRA and the DEMATEL to rank the risk of failures in FMEA, and Chang et al. (2014) integrated the TOPSIS and the DEMATEL methods to analyze the prioritization of failure modes in FMEA.

2.3 Bibliometric Analysis

Bibliometric analysis is a pragmatic research tool used to evaluate a specific field of study. Based on all the relevant papers on FMEA improvements (165 articles), a bibliometric analysis is conducted in this chapter regarding the uncertainty theories adopted in FMEA, the quantity of articles published per year, the journals in which the articles appeared, the application areas of FMEA, the most prolific authors, and the highly cited papers. The results obtained are shown in Figs. 2.1, 2.2, 2.3, 2.4, and 2.5 and Table 2.2. The main purpose of conducting bibliometric analysis is to provide quantitative measures of the analyzed papers. Recent tendencies, distribution of the articles with respect to different categories, and interactions with other fields can give further insights for researchers working in this field.

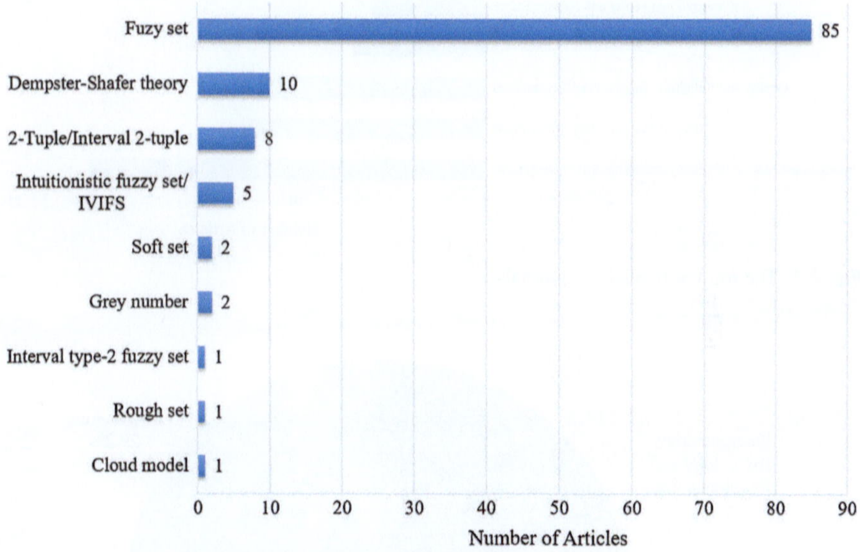

Fig. 2.1 Uncertainty theories used in the reviewed papers

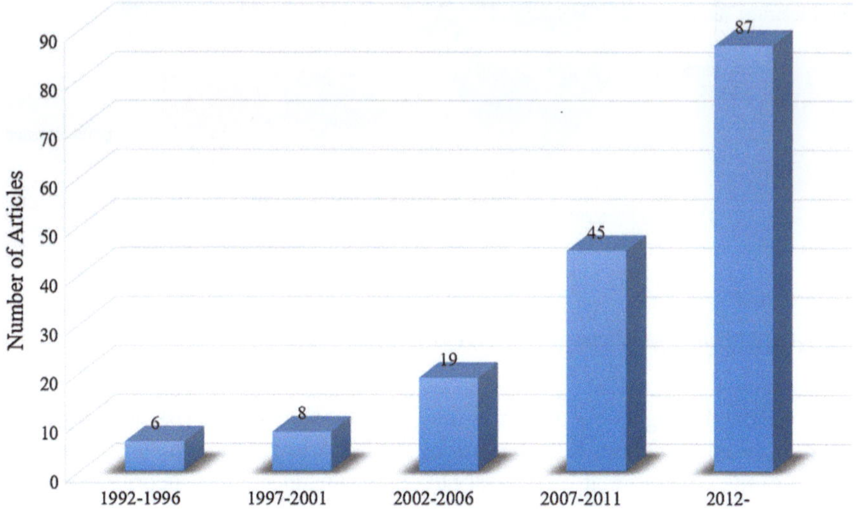

Fig. 2.2 Publishing trend on FMEA improvements

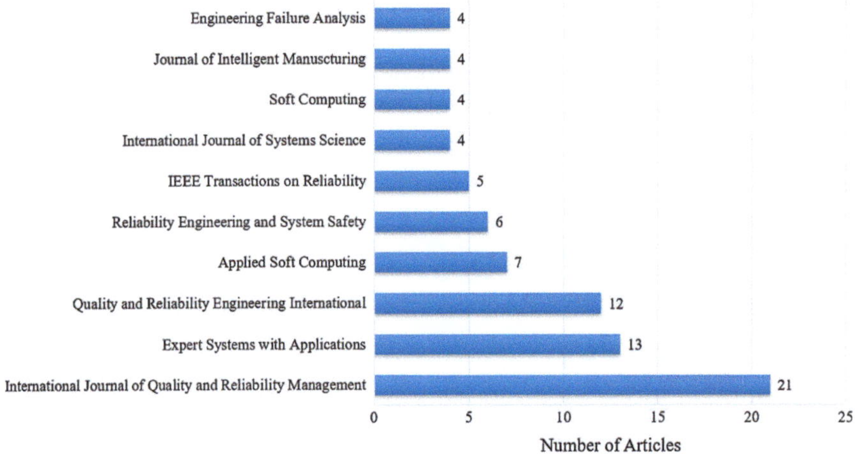

Fig. 2.3 The top ten publishing journals

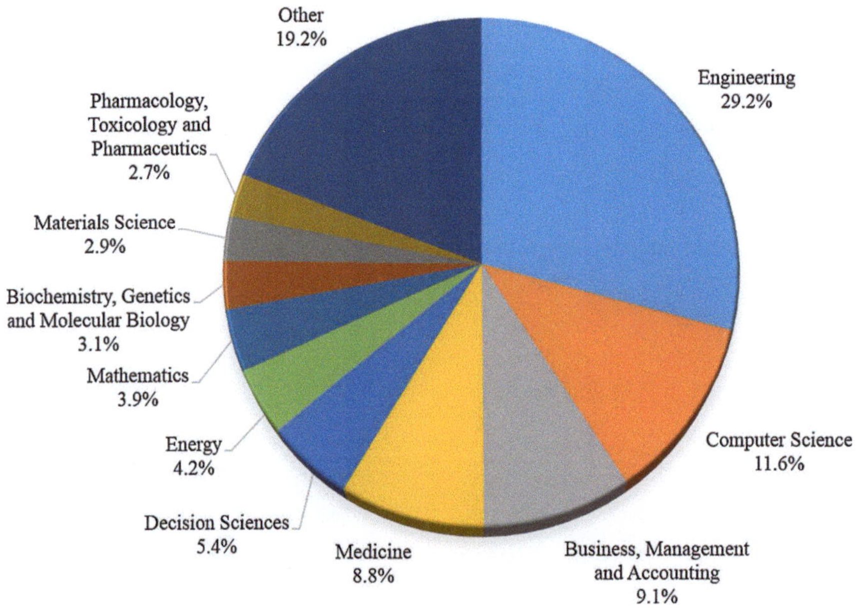

Fig. 2.4 Distribution of articles with respect to application areas

To handle the vagueness of human thought and expression in risk assessments, a lot of uncertainty theories have been used in FMEA to produce more accurate and robust results. Therefore, it is necessary to review the uncertainty theories employed in the collected papers. The number of published papers using every uncertainty method is depicted in Fig. 2.1. As it is shown in Fig. 2.1, fuzzy set is the most

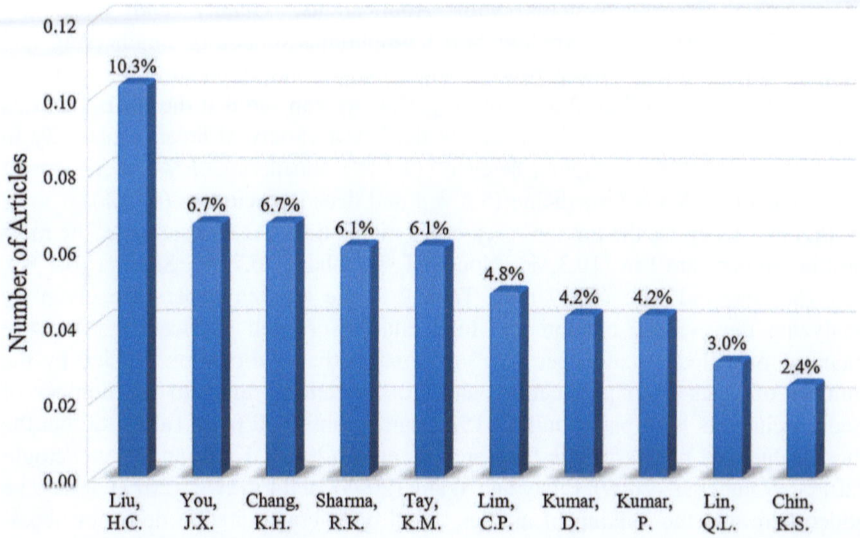

Fig. 2.5 The top ten researchers

Table 2.2 The top 10 papers based on citation measure

Reference	Average citation	Total citation
Liu et al. (2013)	32.50	65
Kutlu and Ekmekcioglu (2012)	29.33	88
Wang et al. (2009)	24.00	144
Pillay and Wang (2003)	18.50	222
Liu et al. (2012)	17.00	51
Chin et al. (2009)	16.33	98
Liu et al. (2011)	14.50	58
Seyed-Hosseini et al. (2006)	14.33	129
Xiao et al. (2011)	14.25	57
Yang et al. (2008)	13.29	93

prevalently applied uncertainty theory for dealing with the ambiguities of FMEA team members' assessments, followed by Dempster–Shafer theory, 2-tuple/interval 2-tuple, and IFS/IVIFS. Recently, the theories such as interval type-2 fuzzy set, grey number, and cloud model are also utilized by researchers to represent and handle vagueness in FMEA.

Then, from Fig. 2.2, one can observe that the number of publications on the modification of FMEA has increased considerably, especially after the year 2007. It can be expected that the studies of improving FMEA will continue to grow at an increased pace in the coming decade. In addition, these articles are mainly published on the journals such as International Journal of Quality and Reliability

Management, Expert Systems with Applications, Quality and Reliability Engineering International, Applied Soft Computing, Reliability Engineering and System Safety, IEEE Transactions on Reliability, and International Journal of Systems Science (See Fig. 2.3). From Fig. 2.4, we can see that the FMEA and its various improvements have been widely used in a variety of areas, practically in engineering (29.2 %), computer science (11.6 %), business (9.1 %), management and accounting (8.8 %), medicine (5.4 %), and decision sciences (4.2 %).

Figure 2.5 depicts the top ten researchers in this field. As one can note, the most prolific authors are Liu (10.3 %), You (6.7 %), Chang (6.7 %), Sharma (6.1 %), Tay (6.1 %), and Lim (4.8 %). In Table 2.2, the top ten papers are given by analyzing the average citation and total citation of each publication. "Average citation" or called "citation per year" is equal to the total citation divided by the number of years from publication, and "total citation" refers to the number of Scopus citations for a paper until 2015. It can be observed from Table 2.2 that the most influenced papers in this filed are Liu et al. (2013), Kutlu and Ekmekçioğlu (2012), Wang et al. (2009), Pillay and Wang (2003), and Liu et al. (2012). It may be added here that the ranking of articles based on average citation does not necessarily match the total citation ranking. For instance, the study by Pillay and Wang (2003) is ranked the fourth in accordance with the average citations, but has the highest total citation value.

References

Abdelgawad M, Fayek AR (2010) Risk management in the construction industry using combined fuzzy FMEA and fuzzy AHP. J Constr Eng Manage 136(9):1028–1036

Adhikary DD, Bose GK, Bose D, Mitra S (2014) Multi criteria FMECA for coal-fired thermal power plants using COPRAS-G. Int J Qual Reliab Manage 31(5):601–614

Braglia M (2000) MAFMA: Multi-attribute failure mode analysis. Int J Qual Reliab Manage 17(9):1017–1033

Braglia M, Frosolini M, Montanari R (2003) Fuzzy TOPSIS approach for failure mode, effects and criticality analysis. Qual Reliab Eng Int 19(5):425–443

Carmignani G (2009) An integrated structural framework to cost-based FMECA: the priority-cost FMECA. Reliab Eng Syst Saf 94(4):861–871

Chang KH (2009) Evaluate the orderings of risk for failure problems using a more general RPN methodology. Microelectron Reliab 49(12):1586–1596

Chang KH (2014) A more general risk assessment methodology using a soft set-based ranking technique. Soft Comput 18(1):169–183

Chang KH (2015) A novel general risk assessment method using the soft TOPSIS approach. J Ind Prod Eng 32(6):408–421

Chang KH (2016) Generalized multi-attribute failure mode analysis. Neurocomputing 175:90–100

Chang KH, Cheng CH (2010) A risk assessment methodology using intuitionistic fuzzy set in FMEA. Int J Syst Sci 41(12):1457–1471

Chang KH, Cheng CH (2011) Evaluating the risk of failure using the fuzzy OWA and DEMATEL method. J Intell Manuf 22(2):113–129

Chang CL, Wei CC, Lee YH (1999) Failure mode and effects analysis using fuzzy method and grey theory. Kybernetes 28(9):1072–1080

Chang CL, Liu PH, Wei CC (2001) Failure mode and effects analysis using grey theory. Integr Manuf Syst 12(3):211–216

Chang KH, Chang YC, Tsai IT (2013) Enhancing FMEA assessment by integrating grey relational analysis and the decision making trial and evaluation laboratory approach. Eng Fail Anal 31:211–224

Chang KH, Chang YC, Lee YT (2014) Integrating TOPSIS and DEMATEL methods to rank the risk of failure of FMEA. Int J Inf Technol Decis Making 13(6):1229–1257

Chin KS, Wang YM, Poon GKK, Yang JB (2009) Failure mode and effects analysis using a group-based evidential reasoning approach. Comput Oper Res 36(6):1768–1779

Du Y, Mo H, Deng X, Sadiq R, Deng Y (2014) A new method in failure mode and effects analysis based on evidential reasoning. Int J Syst Assur Eng Manage 5(1):1–10

Du Y, Lu X, Su X, Hu Y, Deng Y (2016) New failure mode and effects analysis: an evidential downscaling method. Qual Reliab Eng Int 32(2):737–746

Ekmekçioğlu M, Kutlu AC (2012) A fuzzy hybrid approach for fuzzy process FMEA: an application to a spindle manufacturing process. Int J Comput Intell Syst 5(4):611–626

Emovon I, Norman RA, MA J, Pazouki K (2015) An integrated multicriteria decision making methodology using compromise solution methods for prioritising risk of marine machinery systems. Ocean Eng 105:92–103

Franceschini F, Galetto M (2001) A new approach for evaluation of risk priorities of failure modes in FMEA. Int J Prod Res 39(13):2991–3002

Gandhi OP, Agrawal VP (1992) FMEA—A diagraph and matrix approach. Reliab Eng Syst Saf 35(2):147–158

Gargama H, Chaturvedi SK (2011) Criticality assessment models for failure mode effects and criticality analysis using fuzzy logic. IEEE Trans Reliab 60(1):102–110

Geum Y, Cho Y, Park Y (2011) A systematic approach for diagnosing service failure: service-specific FMEA and grey relational analysis approach. Math Comput Model 54(11–12): 3126–3142

Helvacioglu S, Ozen E (2014) Fuzzy based failure modes and effect analysis for yacht system design. Ocean Eng 79:131–141

Hsu CW, Lee WH, Chao B (2013) Materiality analysis model in sustainability reporting-a case study at Lite-On Technology Company. J Clean Prod 57:142–151

Hu AH, Hsu CW, Kuo TC, Wu WC (2009) Risk evaluation of green components to hazardous substance using FMEA and FAHP. Expert Syst Appl 36(3):7142–7147

Ilangkumaran M, Shanmugam P, Sakthivel G, Visagavel K (2014) Failure mode and effect analysis using fuzzy analytic hierarchy process. Int J Prod Quality Manage 14(3):296–313

Jenab K, Dhillon BS (2005) Group-based failure effects analysis. Int J Reliab Qual Saf Eng 12(04):291–307

Kutlu AC, Ekmekçioğlu M (2012) Fuzzy failure modes and effects analysis by using fuzzy TOPSIS-based fuzzy AHP. Expert Syst Appl 39(1):61–67

Li Y, Deng Y, Kang B (2012) A risk assessment methodology based on evaluation of group intuitionistic fuzzy set. J Inf Comput Sci 9(7):1855–1862

Liu HC, Liu L, Bian QH, Lin QL, Dong N, Xu PC (2011) Failure mode and effects analysis using fuzzy evidential reasoning approach and grey theory. Expert Syst Appl 38(4):4403–4415

Liu HC, Liu L, Liu N, Mao LX (2012) Risk evaluation in failure mode and effects analysis with extended VIKOR method under fuzzy environment. Expert Syst Appl 39(17):12926–12934

Liu HC, Liu L, Liu N (2013) Risk evaluation approaches in failure mode and effects analysis: a literature review. Expert Syst Appl 40(2):828–838

Liu HC, Chen YZ, You JX, Li H (2014a) Risk evaluation in failure mode and effects analysis using fuzzy digraph and matrix approach. J Intell Manuf. doi:10.1007/s10845-014-0915-6

Liu HC, Fan XJ, Li P, Chen YZ (2014b) Evaluating the risk of failure modes with extended MULTIMOORA method under fuzzy environment. Eng Appl Artif Intell 34:168–177

Liu HC, Liu L, Li P (2014c) Failure mode and effects analysis using intuitionistic fuzzy hybrid weighted Euclidean distance operator. Int J Syst Sci 45(10):2012–2030

Liu HC, You JX, Fan XJ, Lin QL (2014d) Failure mode and effects analysis using D numbers and grey relational projection method. Expert Syst Appl 41(10):4670–4679
Liu HC, You JX, You XY (2014e) Evaluating the risk of healthcare failure modes using interval 2-tuple hybrid weighted distance measure. Comput Ind Eng 78:249–258
Liu HC, Li P, You JX, Chen YZ (2015a) A novel approach for FMEA: Combination of interval 2-tuple linguistic variables and grey relational analysis. Qual Reliab Eng Int 31(5):761–772
Liu HC, You JX, Ding XF, Su Q (2015b) Improving risk evaluation in FMEA with a hybrid multiple criteria decision making method. Int J Qual Reliab Manage 32(7):763–782
Liu HC, You JX, Lin QL, Li H (2015c) Risk assessment in system FMEA combining fuzzy weighted average with fuzzy decision making trial and evaluation laboratory. Int J Comput Integr Manuf 28(7):701–714
Liu HC, You JX, Shan MM, Shao LN (2015d) Failure mode and effects analysis using intuitionistic fuzzy hybrid TOPSIS approach. Soft Comput 19(4):1085–1098
Liu HC, You JX, You XY, Shan MM (2015e) A novel approach for failure mode and effects analysis using combination weighting and fuzzy VIKOR method. Appl Soft Comput 28:579–588
Liu HC, You JX, Chen SM, Chen YZ (2016a) An integrated failure mode and effect analysis approach for accurate risk assessment under uncertainty. IIE Trans. doi:10.1080/0740817X.2016.1172742
Liu HC, You JX, Li P, Su Q (2016b) Failure mode and effect analysis under uncertainty: an integrated multiple criteria decision making approach. IEEE Trans Reliab (In press)
Lolli F, Ishizaka A, Gamberini R, Rimini B, Messori M (2015) FlowSort-GDSS—A novel group multi-criteria decision support system for sorting problems with application to FMEA. Expert Syst Appl 42(17–18):6342–6349
Moon S, Oh H, Venture J, Kim J, Yoon YJ (2013) Service reliability improvement in manufacturing and operating systems. Int J Precis Eng. Manuf 14(8):1401–1406
Panchal D, Kumar D (2016) Integrated framework for behaviour analysis in a process plant. J Loss Prev Process Ind 40:147–161
Pillay A, Wang J (2003) Modified failure mode and effects analysis using approximate reasoning. Reliab Eng Syst Saf 79(1):69–85
Sachdeva A, Kumar D, Kumar P (2009) Multi-factor failure mode critically analysis using TOPSIS. J Indus Eng Int 5(8):1–9
Safari H, Faraji Z, Majidian S (2016) Identifying and evaluating enterprise architecture risks using FMEA and fuzzy VIKOR. J Intell Manuf 27(2):475–486
Seyed-Hosseini SM, Safaei N, Asgharpour MJ (2006) Reprioritization of failures in a system failure mode and effects analysis by decision making trial and evaluation laboratory technique. Reliab Eng Syst Saf 91(8):872–881
Sharma RK, Sharma P (2012) Integrated framework to optimize RAM and cost decisions in a process plant. J Loss Prev Process Ind 25(6):883–904
Sharma RK, Sharma P (2015) Qualitative and quantitative approaches to analyse reliability of a mechatronic system: A case. J Indus Eng Int 11:253–268
Sharma RK, Kumar D, Kumar P (2007) Modeling system behavior for risk and reliability analysis using KBARM. Qual Reliab Eng Int 23(8):973–998
Sharma RK, Kumar D, Kumar P (2008) Fuzzy modeling of system behavior for risk and reliability analysis. Int J Syst Sci 39(6):563–581
Song W, Ming X, Wu Z, Zhu B (2013) Failure modes and effects analysis using integrated weight-based fuzzy TOPSIS. Int J Comput Integr Manuf 26(12):1172–1186
Song W, Ming X, Wu Z, Zhu B (2014) A rough TOPSIS approach for failure mode and effects analysis in uncertain environments. Qual Reliab Eng Int 30(4):473–486
Tsai TN, Yeh JH (2015) Identification and risk assessment of soldering failure sources using a hybrid failure mode and effect analysis model and a fuzzy inference system. J Intell Fuzzy Syst 28:2772–2784

Vahdani B, Salimi M, Charkhchian M (2015) A new FMEA method by integrating fuzzy belief structure and TOPSIS to improve risk evaluation process. Int J Adv Manuf Technol 77(1):357–368

Wang YM, Chin KS, Poon GKK, Yang JB (2009) Risk evaluation in failure mode and effects analysis using fuzzy weighted geometric mean. Expert Syst Appl 36(2):1195–1207

Wang X, Zhang Y, Shen G (2016) An improved FMECA for feed system of CNC machining center based on ICR and DEMATEL method. Int J Adv Manuf Technol 83:43–54

Xiao NC, Huang HZ, Li YF, He LP, Jin TD (2011) Multiple failure modes analysis and weighted risk priority number evaluation in FMEA. Eng Fail Anal 18(4):1162–1170

Yang Z, Bonsall S, Wang J (2008) Fuzzy rule-based Bayesian reasoning approach for prioritization of failures in FMEA. IEEE Trans Reliab 57(3):517–528

Zammori F, Gabbrielli R (2011) ANP/RPN: A multi criteria evaluation of the risk priority number. Qual Reliab Eng Int 28(1):85–104

Zhao H, You JX, Liu HC (2016) Failure mode and effect analysis using MULTIMOORA method with continuous weighted entropy under interval-valued intuitionistic fuzzy environment. Soft Comput. doi:10.1007/s00500-016-2118-x

Zhou Q, Thai VV (2016) Fuzzy and grey theories in failure mode and effect analysis for tanker equipment failure prediction. Saf Sci 83:74–79

Part II
FMEA Based on Distance-Based MCDM Methods

Chapter 3
FMEA Using Intuitionistic Fuzzy Hybrid Weighted Euclidean Distance Operator

The concept of intuitionistic fuzzy sets (IFSs) is a generalization of fuzzy sets (Zadeh 1965) and was first introduced by Atanassov (1986). The IFS, characterized by membership function, non-membership function, and hesitancy (indeterminancy) function, can depict the fuzzy character of data more comprehensively and is more useful in dealing with vagueness and uncertainty. To overcome the limitations and improve the effectiveness of the traditional FMEA, Liu et al. (2014) developed an efficient and comprehensive risk assessment methodology using intuitionistic fuzzy hybrid weighted Euclidean distance (IFHWED) operator. In this model, the diversified and uncertain assessments given by FMEA team members are treated as linguistic terms expressed in intuitionistic fuzzy numbers (IFNs). The intuitionistic fuzzy weighted averaging (IFWA) operator (Xu 2007a) is utilized to aggregate the FMEA team members' individual assessments for rating the identified failure modes. Finally, the IFHWED operator is applied for the prioritization and selection of failure modes. In particular, both subjective and objective weights of risk factors are taken into account during the risk evaluation process.

3.1 Preliminaries

3.1.1 Intuitionistic Fuzzy Sets

Definition 3.1 (Atanassov 1986) Let X be a fixed set, an IFS A in X is given as following

$$A = \{\langle x, \mu_A(x), v_A(x)\rangle | x \in X\}, \tag{3.1}$$

where $\mu_A(x) : X \to [0, 1]$ and $v_A(x) : X \to [0, 1]$ are membership function and non-membership function, respectively, satisfying $0 \le \mu_A(x) + v_A(x) \le 1, \forall x \in X$.

H.-C. Liu, *FMEA Using Uncertainty Theories and MCDM Methods*,
DOI 10.1007/978-981-10-1466-6_3

The numbers $\mu_A(x)$ and $v_A(x)$ denote, respectively, the degrees of membership and non-membership of the element x to A, for all $x \in X$. In addition, $\pi_A(x) = 1 - \mu_A(x) - v_A(x)$ is called the hesitation degree of $x \in A$, representing the degree of indeterminacy or the degree of hesitancy of x to A. It is obvious that $0 \leq \pi_A(x) \leq 1, \forall x \in X$.

If $\pi_A(x)$ is small, then the value of x is more certain; if $\pi_A(x)$ is greater, then knowledge about x is more uncertain. Additionally, if $\mu_A(x)$ and $v_A(x)$ are both continuous functions, the IFS explanation of a real number R can be shown in Fig. 3.1. Obviously, when $\mu_A(x) = 1 - v_A(x)$, for all elements of the universe, the IFS reduces to a traditional fuzzy set; when $\mu_A(x) = 1 - v_A(x) = 1$ or $\mu_A(x) = 1 - v_A(x) = 0$, the IFS regresses to a crisp set. Hence, crisp sets and fuzzy sets can be viewed as special cases of IFSs.

For an IFS, the pair $(\mu_A(x), v_A(x))$ is called an intuitionistic fuzzy number (IFN) (Xu 2007a) and each IFN can be simply denoted as $\alpha = (\mu_\alpha, v_\alpha)$, where $\mu_\alpha \in [0,1], v_\alpha \in [0,1]$ and $\mu_\alpha + v_\alpha \leq 1$. For an IFN $\alpha = (\mu_\alpha, v_\alpha)$, if the value μ_α gets bigger and the value v_α gets smaller, then the IFN $\alpha = (\mu_\alpha, v_\alpha)$ gets greater. Obviously, $\alpha^+ = (1,0)$ and $\alpha^- = (0,1)$ are the largest and the smallest IFNs, respectively. In addition, $S(\alpha) = \mu_\alpha - v_\alpha$ and $H(\alpha) = \mu_\alpha + v_\alpha$ are called the score and accuracy degrees of α, respectively.

Definition 3.2 For any three IFNs $\alpha_1 = (\mu_{\alpha_1}, v_{\alpha_1}), \alpha_2 = (\mu_{\alpha_2}, v_{\alpha_2})$, and $\alpha = (\mu_\alpha, v_\alpha)$, the following operational laws are introduced (Xu and Yager 2006; Xu 2007a):

(1) $\alpha_1 + \alpha_2 = (\mu_{\alpha_1} + \mu_{\alpha_2} - \mu_{\alpha_1}\mu_{\alpha_2}, v_{\alpha_1}v_{\alpha_2})$,

(2) $\alpha_1 \times \alpha_2 = (\mu_{\alpha_1}\mu_{\alpha_2}, v_{\alpha_1} + v_{\alpha_2} - v_{\alpha_1}v_{\alpha_2})$,

(3) $\lambda\alpha = \left(1 - (1 - \mu_\alpha)^\lambda, v_\alpha^\lambda\right), \quad \lambda > 0$,

(4) $\alpha^\lambda = \left(\mu_\alpha^\lambda, 1 - (1 - v_\alpha)^\lambda\right), \quad \lambda > 0$.

Definition 3.3 For comparing any two IFNs α_1 and α_2, the following method was proposed based on the score function and the accuracy function (Xu and Yager 2006; Xu 2007a):

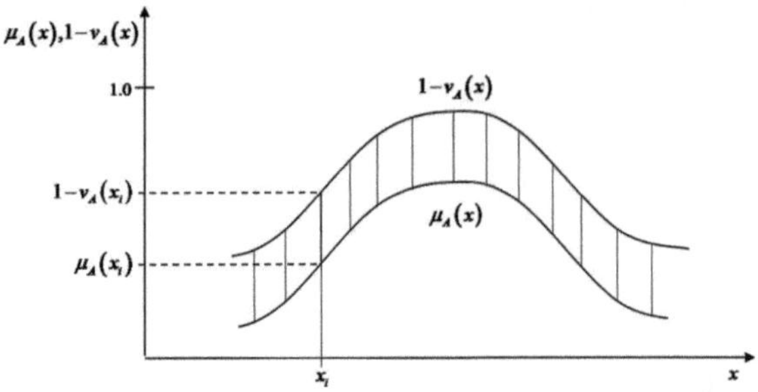

Fig. 3.1 Illustration of intuitionistic fuzzy sets

(1) If $S(\alpha_1) < S(\alpha_2)$, then $\alpha_1 < \alpha_2$;
(2) If $S(\alpha_1) = S(\alpha_2)$, and

  (a) If $H(\alpha_1) < H(\alpha_2)$, then $\alpha_1 < \alpha_2$;
  (b) If $H(\alpha_1) = H(\alpha_2)$, then $\alpha_1 = \alpha_2$.

In order to measure the deviation between any two IFNs, Xu (2007b, 2010) defined the following distance.

Definition 3.4 Let $\alpha_1 = (\mu_{\alpha_1}, v_{\alpha_1})$ and $\alpha_2 = (\mu_{\alpha_2}, v_{\alpha_2})$ be two IFNs, then

$$d_{\mathrm{IFD}}(\alpha_1, \alpha_2) = |\alpha_1 - \alpha_2| = \frac{1}{2}\left(\left|\mu_{\alpha_1} - \mu_{\alpha_2}\right| + \left|v_{\alpha_1} - v_{\alpha_2}\right|\right) \tag{3.2}$$

is called the intuitionistic fuzzy distance (IFD) between α_1 and α_2.

3.1.2 The IFWA Operator

Following the conceptions and operations of IFNs, let Ω be the set of all IFNs, Xu (2007a) introduced the intuitionistic fuzzy weighted averaging (IFWA) operator as follows.

Definition 3.5 Let $\alpha_i = (\mu_{\alpha_i}, v_{\alpha_i}), i = 1, 2, \ldots, n$ be a collection of IFNs and let $\mathrm{IFWA} : \Omega^n \to \Omega$, if

$$\begin{aligned} \mathrm{IFWA}(\alpha_1, \alpha_2, \ldots, \alpha_n) &= w_1\alpha_1 + w_2\alpha_2 + \cdots + w_n\alpha_n \\ &= \left(1 - \prod_{i=1}^{n}\left(1 - \mu_{\alpha_i}\right)^{w_i}, \prod_{i=1}^{n}\left(v_{\alpha_i}\right)^{w_i}\right), \end{aligned} \tag{3.3}$$

then IFWA is called the intuitionistic fuzzy weighted averaging operator of dimension n, where $w = (w_1, w_2, \ldots, w_n)^T$ is the weight vector of $\alpha_i (i = 1, 2, \ldots, n)$, with $w_i \in [0, 1]$ and $\Sigma_{i=1}^n w_i = 1$.

3.1.3 The OWA Operator

The ordered weighted averaging (OWA) operator introduced by Yager (1988) provides a parameterized family of aggregation operators that include the maximum, the minimum, and the average criteria. Its prominent advantage is that the input data are rearranged in descending order, and the weights associated with the

OWA operator are the weights of the ordered positions of the input data rather than the weights of the input data. The OWA operator can be defined as follows.

Definition 3.6 An OWA operator of dimension n is a mapping OWA : $R^n \rightarrow R$ that has an associated weighting vector $\omega = (\omega_1, \omega_2, \ldots, \omega_n)^T$, with $\omega_j \in [0, 1]$ and $\Sigma_{j=1}^{n} \omega_j = 1$, such that

$$\text{OWA}(a_1, a_2, \ldots, a_n) = \sum_{j=1}^{n} \omega_j b_j, \tag{3.4}$$

where b_j is the jth largest of the a_i. The OWA operator is commutative, monotonic, bounded, and idempotent. From a generalized perspective of the reordering step, it is possible to distinguish between the descending OWA operator and the ascending OWA operator.

Determining the OWA weights is one key point in the OWA operator, and a number of methods have been developed for obtaining its associated weights (Yager 1988; Fullér and Majlender 2001; Xu 2005; Liu et al. 2015). To relieve the influence of unfair arguments on the decision results, Xu (2005) suggested a normal distribution-based method to generate the weights of the OWA operator. By using this method, the associated weighting vector is obtained by

$$\omega_i = \frac{1}{\sqrt{2\pi}\sigma_n} e^{-\left[(i-\mu_n)^2/2\sigma_n^2\right]} = \frac{e^{-\left[(i-\mu_n)^2/2\sigma_n^2\right]}}{\sum_{i=1}^{n} e^{-\left[(i-\mu_n)^2/2\sigma_n^2\right]}}, \quad i = 1, 2, \ldots, n, \tag{3.5}$$

where $\omega = (\omega_1, \omega_2, \ldots, \omega_n)^T$ is the weigh vector of the OWA operator, μ_n is the mean of the collection of $1, 2, \ldots, n$, and $\sigma_n (\sigma_n > 0)$ is the standard deviation of the collection of $1, 2, \ldots, n.\mu_n$, and σ_n can be obtained by the following formulas, respectively:

$$\mu_n = \frac{1}{n} \frac{n(1+n)}{2} = \frac{1+n}{2}, \tag{3.6}$$

$$\sigma_n = \left(\frac{1}{n} \sum_{i=1}^{n} (i - \mu_n)^2 \right)^{1/2}. \tag{3.7}$$

3.1.4 The IFWED and the IFOWED Operators

Zeng and Su (2011) developed an intuitionistic fuzzy ordered weighted distance (IFOWD) operator. The main advantage of the IFOWD operator is that it can alleviate the influence of unduly large (or small) deviations on the aggregation results by assigning them low weights. The intuitionistic fuzzy weighted Euclidean distance (IFWED) and the intuitionistic fuzzy ordered weighted Euclidean distance (IFOWED) operators are special cases of the IFOWD operator. For two sets of IFNs $\tilde{A} = (\alpha_1, \alpha_2, \ldots, \alpha_n)$ and $\tilde{B} = (\beta_1, \beta_2, \ldots, \beta_n)$, they can be defined as follows.

Definition 3.7 An IFWED operator of dimension n is a mapping IFWED : $\Omega^n \times \Omega^n \to R$ that has an associated weighting vector $w = (w_1, w_2, \ldots, w_n)^T$, with $w_i \in [0, 1]$ and $\Sigma_{i=1}^n w_i = 1$, according to the following formula:

$$\mathrm{IFWED}(\tilde{A}, \tilde{B}) = \left(\sum_{i=1}^{n} w_i (d_{\mathrm{IFD}}(\alpha_i, \beta_i))^2 \right)^{1/2}. \tag{3.8}$$

Definition 3.8 An IFOWED operator of dimension n is a mapping IFOWED : $\Omega^n \times \Omega^n \to R$ that has an associated weighting vector $\omega = (\omega_1, \omega_2, \ldots, \omega_n)^T$, with $\omega_j \in [0, 1]$ and $\Sigma_{j=1}^n \omega_j = 1$, according to the following formula:

$$\mathrm{IFOWED}(\tilde{A}, \tilde{B}) = \left(\sum_{j=1}^{n} \omega_j \left(d_{\mathrm{IFD}} \left(\alpha_{\sigma(j)}, \beta_{\sigma(j)} \right) \right)^2 \right)^{1/2}, \tag{3.9}$$

where $(\sigma(1), \sigma(2), \ldots, \sigma(n))$ is any permutation of $(1, 2, \ldots, n)$, such that $d_{\mathrm{IFD}}\left(\alpha_{\sigma(j-1)}, \beta_{\sigma(j-1)}\right) \geq d_{\mathrm{IFD}}\left(\alpha_{\sigma(j)}, \beta_{\sigma(j)}\right), j = 1, 2, \ldots, n$.

3.1.5 The IFHWED Operator

Liu et al. (2014) presented a new approach to unify the IFWED operator with the IFOWED operator when the decision information is provided with IFNs, named the intuitionistic fuzzy hybrid weighted Euclidean distance (IFHWED) operator. Its main advantage is that it can unify both concepts considering the degree of importance that each one has in the aggregation process. It can be defined as follows.

Definition 3.9 An IFHWED operator of dimension n is a mapping IFHWED : $\Omega^n \times \Omega^n \to R$ that has an associated weighting vector $w = (w_1, w_2, \ldots, w_n)^T$, with $w_i \in [0, 1]$ and $\Sigma_{i=1}^n w_i = 1$, and a weighting vector $\omega = (\omega_1, \omega_2, \ldots, \omega_n)^T$, with $\omega_j \in [0, 1]$ and $\Sigma_{j=1}^n \omega_j = 1$, according to the following formula:

$$\begin{aligned}\mathrm{IFHWED}(\tilde{A},\tilde{B}) &= \varphi\left(\sum_{i=1}^{n} w_i(d_{\mathrm{IFD}}(\alpha_i,\beta_i))^2\right)^{1/2} \\ &\quad + (1-\varphi)\left(\sum_{j=1}^{n} \omega_j\left(d_{\mathrm{IFD}}\left(\alpha_{\sigma(j)},\beta_{\sigma(j)}\right)\right)^2\right)^{1/2},\end{aligned} \tag{3.10}$$

where $(\sigma(1),\sigma(2),\ldots,\sigma(n))$ is any permutation of $(1,2,\ldots,n)$, such that $d_{\mathrm{IFD}}\left(\alpha_{\sigma(j-1)},\beta_{\sigma(j-1)}\right) \geq d_{\mathrm{IFD}}\left(\alpha_{\sigma(j)},\beta_{\sigma(j)}\right), j=1,2,\ldots,n$, and $\varphi \in [0,1]$. As we can see, if $\varphi = 1$, we get the IFWED operator and if $\varphi = 0$, the IFOWED operator.

3.2 The Proposed FMEA Approach

In the real-life world, due to the increasing complexity of the assessed systems and the lack of knowledge or data about the problem domain, the risk factors are not easy to be precisely evaluated. As such, in this chapter, we choose linguistic terms for the assessment of risk factors and the individual evaluation grade is defined as an IFN. Table 3.1 shows the linguistic terms and their IFNs used for evaluating risk factors. Moreover, the great majority of improved FMEA methods only consider the subjective or objective weights of risk factors separately. To overcome this drawback, both subjective and objective weights of risk factors are considered in the proposed FMEA. The subjective weights of risk factors are assessed by FMEA team members using the linguistic terms as provided in Table 3.2. The objective risk factor weights are determined by the ordered weights of risk factors, which are derived by the normal distribution-based method (Xu 2005).

The flowchart in Fig. 3.2 shows the proposed approach to rank the failure modes, which are identified in the FMEA process. Three key steps are included in the proposed approach: aggregation, calculation, and ranking. The FMEA team gives their individual judgments on failure modes by using linguistic terms defined

Table 3.1 Linguistic terms for rating failure modes

Linguistic terms	IFNs
Extremely low (EL)	(0.10, 0.90)
Very low (VL)	(0.25, 0.70)
Low (L)	(0.30, 0.60)
Medium low (ML)	(0.40, 0.50)
Medium (M)	(0.50, 0.50)
Medium high (MH)	(0.60, 0.30)
High (H)	(0.70, 0.20)
Very high (VH)	(0.75, 0.20)
Extremely high (EH)	(0.90, 0.10)

Table 3.2 Linguistic terms for rating subjective weights of risk factors

Linguistic terms	IFNs
Very low (VL)	(0.10, 0.85)
Low (L)	(0.25, 0.70)
Moderate (M)	(0.50, 0.50)
High (H)	(0.75, 0.20)
Very high (VH)	(0.90, 0.05)

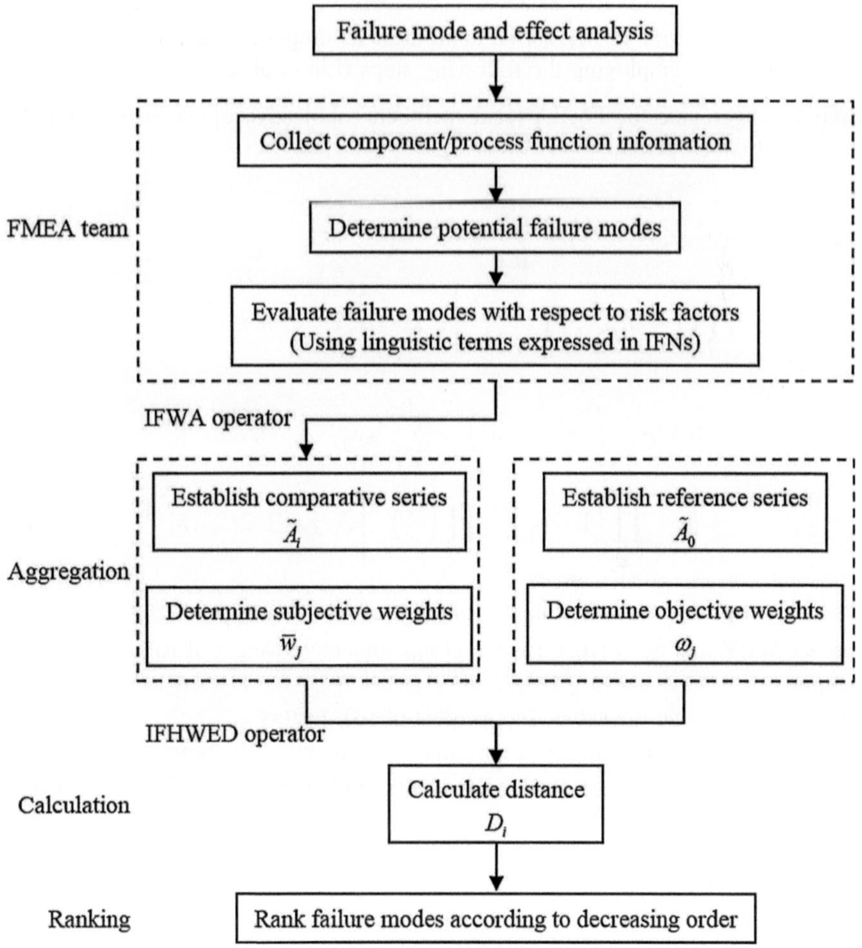

Fig. 3.2 Flowchart of the proposed FMEA approach (Liu et al. 2014)

by IFNs. The IFWA operator is cited for aggregating these judgments in order to form a consensus group judgment. Incorporated with the subjective and ordered weights, the IFHWED operator is used for calculating the distance between the reference series with the aggregated results. Finally, the risk ranking of failure modes can be yielded according to the results obtained in the previous step.

Suppose there are l cross-functional members, $TM_k(k = 1, \ldots, l)$, in a FMEA team responsible for the assessment of m failure modes, $FM_i(i = 1, \ldots, m)$, with respect to n risk factors, RF_j $(j = 1, \ldots, n)$. Each team member TM_k is given a weight $\lambda_k > 0(k = 1, \ldots, l)$ satisfying $\Sigma_{k=1}^{l} \lambda_k = 1$ to reflect his/her relative importance in the FMEA team. Let $\alpha_{ij}^k = \left(\mu_{ij}^k, v_{ij}^k\right)$ be the IFN provided by TM_k on the assessment of FM_i with respect to RF_j, and $w_j^k = \left(\mu_j^k, v_j^k\right)$ be the subjective weight of RF_j given by TM_k. Based upon these assumptions, the failure modes can be prioritized by employing the following steps (Liu et al. 2014):

Step 1. Aggregate the FMEA team members' subjective opinions by using the IFWA operator

$$\alpha_{ij} = \text{IFWA}\left(\alpha_{ij}^1, \alpha_{ij}^2, \ldots, \alpha_{ij}^l\right) = \sum_{k=1}^{l} \lambda_k \alpha_{ij}^k$$
$$= \left[1 - \prod_{k=1}^{l}\left(1 - \mu_{ij}^k\right)^{\lambda_k}, \prod_{k=1}^{l}\left(v_{ij}^k\right)^{\lambda_k}\right], \quad i = 1, 2, \ldots, m; j = 1, 2, \ldots, n, \tag{3.11}$$

$$w_j = \text{IFWA}\left(w_j^1, w_j^2, \ldots, w_j^l\right) = \sum_{k=1}^{l} \lambda_k w_j^k$$
$$= \left[1 - \prod_{k=1}^{l}\left(1 - \mu_j^k\right)^{\lambda_k}, \prod_{k=1}^{l}\left(v_j^k\right)^{\lambda_k}\right], \quad j = 1, 2, \ldots, n, \tag{3.12}$$

where $\alpha_{ij} = \left(\mu_{ij}, v_{ij}\right)$ is the group assessment of the l team members for FM_i with respect to RF_j and $w_j = \left(\mu_j, v_j\right)$ is the group subjective weight of RF_j of the l team members.

Step 2. Determine the subjective weights of risk factors

Based on the group subjective weights $w_j = \left(\mu_j, v_j\right)$, determined by Eq. (3.12), the subjective weight of risk factor RF_j can be normalized using Eq. (3.14) (Boran et al. 2009).

$$\bar{w}_j = \frac{\mu_j + \pi_j\left(\frac{\mu_j}{\mu_j + v_j}\right)}{\sum_{j=1}^{n}\left(\mu_j + \pi_j\left(\frac{\mu_j}{\mu_j + v_j}\right)\right)}, \quad j = 1, \ldots, n, \tag{3.13}$$

where $\pi_j = 1 - \mu_j - v_j$ is hesitation degree and $\Sigma_{j=1}^{n} \bar{w}_j = 1$.

Step 3. Determine the objective weights of risk factors

The normal distribution-based method suggested by Xu (2005) is employed here to calculate the objective weights of risk factors. As a result, the associated weighting vector can be obtained by Eqs. (3.5)–(3.7). For example, if $n = 3$, by Eqs. (3.6) and (3.7), we get $\mu_3 = 2$ and $\sigma_3 = \sqrt{2/3}$; then from Eq. (3.5), we can get the objective weight vector as $\omega = (0.243, 0.514, 0.243)^T$.

Step 4. Establish the reference series for the risk factors

The reference series for the risk factors should be the optimal level of all risk factors for the failure modes in FMEA. When conducting FMEA, the smaller the score, the less the risk; therefore, the minimum value $\alpha^- = (0, 1)$ can be used as the reference series, which is expressed as follows:

$$\tilde{A}_0 = [\alpha_{01}, \alpha_{02}, \ldots, \alpha_{0n}] = [\alpha^-, \alpha^-, \ldots, \alpha^-]. \tag{3.14}$$

Step 5. Calculate the distances between the reference series with the aggregated results by using the IFHWED operator

A comparative series with n components or risk factors can be expressed as $\tilde{A}_i = [\alpha_{i1}, \alpha_{i2}, \ldots, \alpha_{in}]$, where $\alpha_{ij}(j = 1, \ldots, n)$ are obtained by Eq. (3.11). Then, the IFHWED between the comparative and the reference series, D_i, can be calculated as follows:

$$\begin{aligned} D_i &= \text{IFHWED}(\tilde{A}_i, \tilde{A}_0) \\ &= \varphi\left(\sum_{j=1}^{n} \bar{w}_j\left(d_{\text{IFD}}\left(\alpha_{ij}, \alpha_{0j}\right)\right)^2\right)^{1/2} + (1 - \varphi)\left(\sum_{j=1}^{n} \omega_j\left(d_{\text{IFD}}\left(\alpha_{i\sigma(j)}, \alpha_{0\sigma(j)}\right)\right)^2\right)^{1/2}, \\ & i = 1, \ldots m, \end{aligned} \tag{3.15}$$

where $(\sigma(1), \ldots, \sigma(n))$ is any permutation of $(1, \ldots, n)$, such that $d_{\text{IFD}}\left(\alpha_{i\sigma(j-1)}, \alpha_{0\sigma(j-1)}\right) \geq d_{\text{IFD}}\left(\alpha_{i\sigma(j)}, \alpha_{0\sigma(j)}\right), j = 1, \ldots, n$, and $\varphi \in [0, 1]$.

Step 6. Rank all the failure modes

The ranking order of all the failure modes can be derived according to the decreasing order of their IFHWEDs. The bigger the distance, the higher the overall risk of the failure mode.

In the above computations, the relative weights of FMEA team members are assumed to be crisp values because they are relatively easier to be determined. In real-world applications, they can be acquired by using direct grade, point allocation, eigenvector method, or Delphi method, etc., together with the team members' experience and domain knowledge (Chin et al. 2009). In addition, if necessary, the importance degree for each of the team members can also be assessed using the linguistic terms in Table 3.2 and calculated by using Eq. (3.13).

3.3 An Illustrative Example

3.3.1 Implementation

In this section, an illustrative example of developing new horizontal directional drilling (HDD) machine is presented in order to demonstrate the procedure that is proposed in this chapter. The FMEA example is adapted from (Zhang and Chu 2011; Liu et al. 2014). The HDD machine, as the key equipment for trenchless construction, is a typical complex product that consists of several multi-disciplinary subsystems, such as mechanism, hydraulic system, electric system, and engine system.

A FMEA team consisting of five cross-functional members identifies nine potential failure modes in the HDD machine development and needs to prioritize them in terms of their risk factors such as *O*, *S* and *D* so that high risky failure modes can be corrected with top priorities. The failure modes are identified as gear abrasion of dynamic head (FM1), action invalidation of force motor (FM2), non-normal friction of pedrail (FM3), leak of hydraulic system (FM4), abrasion of feed mechanism (FM5), unexpected halt of engine (FM6), cavitation erosion of hydraulic pump (FM7), failures of hydraulic system induced by hydraulic oil pollution (FM8), and nozzle choking of aiguilles (FM9). Due to the difficulty in precisely assessing the risk factors and their importance weights, the FMEA team members are assumed to evaluate them by employing the linguistic terms expressed in IFNs in Tables 3.1 and 3.2. The assessment information of the nine failure modes on each risk factor and the risk factor weights provided by the five team members can be seen in Table 3.3. The five team members from different departments, e.g., design, manufacturing, and technical service, are assumed to be of different importance because of their different domain knowledge and expertise. To reflect their differences in performing FMEA, the five team members are assigned the following relative weights: 0.15, 0.20, 0.25, 0.10, and 0.30.

After quantifying by corresponding IFNs, the FMEA team members' individual assessments are aggregated into group assessments by utilizing Eqs. (3.11) and (3.12). The results so obtained are presented in Table 3.4. After the determination of the group weights, using Eq. (3.13), the subjective weight vector for the risk factors is obtained as $\bar{w} = (0.362, 0.485, 0.153)^T$. On the other side, the ordered weight vector for the risk factors is derived as $\omega = (0.243, 0.514, 0.243)^T$ by the normal distribution-based method. Next, by using the subjective and the ordered weight vectors of the risk factors, and the group assessments of the FMEA team, the IFHWED is calculated via Eq. (3.15) for each of the failure modes identified in the FMEA. In this example, the parameter φ is assumed to be 0.6 and the reference series $\tilde{A}_0 = [(0,1),(0,1),\ldots,(0,1)]$. The results are summarized in Table 3.5. The identified failure modes in the FMEA are ranked according to the decreasing order of their IFHWEDs. This entails that the failure mode with the largest distance gets the highest priority for attention. For the example in Table 3.5, FM7 would be at the top of the list for priority for attention, followed by FM2, FM8, FM6, FM3, FM5, FM1, FM9, and FM4.

Table 3.3 Assessed information on failure modes from the FMEA team (Liu et al. 2014)

Risk factors		*O*					*S*					*D*				
Team members		TM1	TM2	TM3	TM4	TM5	TM1	TM2	TM3	TM4	TM5	TM1	TM2	TM3	TM4	TM5
Factor weight		H	M	H	M	M	H	VH	H	VH	H	VL	L	L	M	L
Failure modes	FM1	M	M	M	MH	M	ML	MH	ML	M	M	M	L	ML	ML	ML
	FM2	M	MH	H	MH	MH	MH	MH	H	M	H	MH	MH	ML	MH	M
	FM3	ML	ML	L	M	M	MH	M	H	MH	M	H	H	H	MH	MH
	FM4	L	M	L	M	M	M	MH	ML	M	M	VL	ML	VL	L	VL
	FM5	M	ML	M	M	M	M	H	MH	MH	M	L	ML	VL	L	VL
	FM6	H	MH	M	MH	M	H	H	MH	MH	H	L	M	L	EL	VL
	FM7	VH	MH	VH	MH	H	MH	MH	H	MH	MH	MH	M	MH	M	M
	FM8	ML	M	L	ML	ML	EH	H	EH	H	VH	ML	ML	L	L	L
	FM9	M	M	ML	M	M	ML	M	ML	ML	M	M	M	M	ML	ML

Table 3.4 Aggregated assessed information for the failure modes and the subjective weights of risk factors (Liu et al. 2014)

Failure modes	*O*	*S*	*D*
FM1	(0.511, 0.475)	(0.486, 0.451)	(0.398, 0.519)
FM2	(0.615, 0.293)	(0.651, 0.253)	(0.527, 0.397)
FM3	(0.420, 0.523)	(0.584, 0.350)	(0.663, 0.235)
FM4	(0.428, 0.538)	(0.500, 0.451)	(0.288, 0.644)
FM5	(0.481, 0.500)	(0.582, 0.348)	(0.295, 0.630)
FM6	(0.567, 0.374)	(0.668, 0.230)	(0.315, 0.631)
FM7	(0.696, 0.226)	(0.628, 0.271)	(0.543, 0.408)
FM8	(0.399, 0.523)	(0.817, 0.152)	(0.337, 0.563)
FM9	(0.477, 0.500)	(0.452, 0.500)	(0.462, 0.500)
w_j	(0.621, 0.347)	(0.810, 0.132)	(0.260, 0.697)

Table 3.5 IFD and IFHWED for the nine failure modes and risk ranking (Liu et al. 2014)

Failure modes	*O*	*S*	*D*	IFHWED	Ranking
FM1	0.518	0.518	0.440	0.504	7
FM2	0.661	0.699	0.565	0.659	2
FM3	0.449	0.617	0.714	0.590	5
FM4	0.445	0.525	0.322	0.458	9
FM5	0.491	0.617	0.333	0.520	6
FM6	0.597	0.719	0.342	0.610	4
FM7	0.735	0.679	0.568	0.678	1
FM8	0.438	0.833	0.387	0.613	3
FM9	0.489	0.476	0.481	0.481	8

3.3.2 *Sensitivity Analysis*

A sensitivity analysis by changing the parameter φ is calculated according to the information shown in Table 3.5. The risk priority rankings of the nine failure modes under different φ values are represented in Table 3.6. From Table 3.6, the following findings have been discovered (Liu et al. 2014):

- Failure modes can also be prioritized when only the subjective or the objective weights of risk factors are considered, but this will result in biased or even misleading rankings. So, setting an appropriate φ according to actual situations and experts' opinions is of significance and benefit to the risk prioritization of failure modes and the following corrective actions.
- The final ranking orders can be certainly influenced by the setting of weight restrictions φ. With the changing of the weight restriction from 0 to 1, the risk priority rankings of four of nine failure modes (44.4 %) are different, such as FM1, FM3, FM5, and FM8. In this sense, the differences between the rankings under different weight restrictions are very big.

Table 3.6 Risk priority rankings under different weight restrictions (Liu et al. 2014)

Failure modes	Ranking					
	$\varphi = 0$	$\varphi = 0.2$	$\varphi = 0.4$	$\varphi = 0.6$	$\varphi = 0.8$	$\varphi = 1$
FM1	6	7	7	7	7	7
FM2	2	2	2	2	2	2
FM3	3	3	4	5	5	5
FM4	9	9	9	9	9	9
FM5	7	6	6	6	6	6
FM6	4	4	3	4	4	4
FM7	1	1	1	1	1	1
FM8	5	5	5	3	3	3
FM9	8	8	8	8	8	8

- The priority rankings of the nine failure modes for $\varphi \geq 0.6$ are not changed. This is because, except the weight restriction, the risk priorities of failure modes are also affected by the subjective weights, the objective weights, and the d_{IFD} values of risk factors.

From the sensitivity analysis, we can see that proper selection of φ plays an important role in the criticality analysis because it may affect the final rankings of the failure modes. Normally, the weight restriction φ is determined according to the following ways: First, it can be determined by decision makers by referring to historical data if they have conducted a similar FMEA analysis before. Second, it can be assigned by decision makers on the basis of actual situations. For example, the parameter φ can be given a lower value if it is difficult or undesirable to get subjective weights, i.e., when experts are difficult to reach an agreement on the relative importance of risk factors or suitable experts are not available. Third, the weight restriction can be generated with the questionnaire answered by domain experts. If he/she is more confident about his/her judgments with respect to the subjective risk factor weights, he/she can give a higher value to φ; otherwise, the weight restriction should be a small number, such as less than 0.5. In addition, the two kinds of weights can be assumed to be equally important and the parameter φ can be set to 0.5 in the worst cases.

3.3.3 Comparisons and Discussion

To illustrate the effectiveness of the proposed FMEA approach, we used the above case study to analyze some comparable methods, which include the fuzzy FMEA (Pillay and Wang 2003), the OWA-based FMEA (Chang and Cheng 2011), and the intuitionistic fuzzy FMEA (IF-FMEA) (Chang and Cheng 2010). Table 3.7 exhibits the ranking results of the nine failure modes as obtained using these approaches.

Table 3.7 Ranking comparison (Liu et al. 2014)

Failure modes	Proposed approach	Fuzzy FMEA	OWA-based FMEA	IF-FMEA
FM1	7	6	8	6
FM2	2	1	1	2
FM3	5	5	3	3
FM4	9	9	9	9
FM5	6	7	7	8
FM6	4	2	5	4
FM7	1	1	1	1
FM8	3	2	4	5
FM9	8	8	6	7

Based on the results in Table 3.7, the advantages that the proposed method has over other methods can be identified.

First, from Table 3.7, we can see that the ranking orders of FM2 and FM7, FM6, and FM8 are the same and FM5 is ranked behind FM1 when the fuzzy FMEA is applied. However, in the proposed method, FM2 and FM6 are successfully distinguished from FM7 and FM8, respectively. In addition, the results of the proposed method show that FM5 has a high priority compared to FM1. The main reasons for these differences may be as follows (Wang et al. 2009; Liu et al. 2013, 2014): (1) The fuzzy if–then rules with the same consequence but different antecedents are unable to be distinguished from one another by using fuzzy FMEA. (2) Reduced rules will be incomplete if they are not reduced from a complete if–then rule base. Any inference from an incomplete rule base will be biased or even wrong because some knowledge cannot be learned from such an incomplete rule base. (3) The use of fuzzy if–then rules has no way to incorporate the relative importance of risk factors into the fuzzy inference system.

Second, there is much difference between the two sets of risk priority rankings produced by the proposed FMEA and the OWA-based FMEA. The explanations of the inconsistent ranking results are summarized as follows (Liu et al. 2014): (1) When each cause of failure is assigned to only one failure mode, the risk ranking orders obtained by the OWA-based FMEA corresponds with the ones obtained by the OWA operator. However, the mathematical formula for calculating the severity of influences among failure modes is questionable. Different combinations of O, S, and D may produce exactly the same value, but their risk levels may be totally different. In contrast, the basic principle of our proposed FMEA is that the high risky failure mode should have the "farthest distance" from the reference series. (2) It will lose some information when use the maximum membership degree to calculate the aggregated value by the OWA weights. Consequently, the information in other degrees of membership is lost, which implies a lack of precision in the final results. (3) The weighing calculation methods in the two methods are different. The OWA-based FMEA used the method of Lagrange multipliers (Fullér and Majlender 2001) to determine the OWA weights. Although this weighting method can reflect the aggregate situation during the aggregation process, it is hard to determine an appropriate situation parameter in practice. In the proposed FMEA, however, the ordered weights for the risk factors were derived by the normal distribution-based method (Xu 2005), which can relieve the influence of unfair arguments on the decision results by weighting these arguments with small values. (4) The OWA-based FMEA ignores the subjective weights of risk factors, which may cause biased ranking results.

Third, the prioritization of failure modes obtained from the IF-FMEA is different from the ranking produced from the proposed risk priority model, which can be explained by the following reasons (Liu et al. 2014): (1) When each cause of failure is assigned to only one potential failure mode, the risk ranking orders obtained by the IF-FMEA and the weighted RPN are the same. As a result, the IF-FMEA will have similar shortcomings as the conventional RPN method. (2) In the IF-FMEA, the triangular IFSs were reduced to exact values through a defuzzification method modified by the authors. These reductions lead to miss some original decision information. On the contrary, the defuzzification step is avoided in the proposed approach. (3) The objective risk factor weights are not considered during the risk analysis of the IF-FMEA, which may cause biased conclusions.

The analysis of the results produced by the fuzzy FMEA, the OWA-based FMEA, and the IF-FMEA show that a more accurate, reasonable risk assessment can be achieved by applying the IFHWED operator to FMEA. Moreover, according to the domain experts, the proposed risk priority model with subjective and objective weights is more suitable for the risk evaluation problem examined and can find the most critical failures effectively.

References

Atanassov K (1986) Intuitionistic fuzzy sets. Fuzzy Sets Syst 20(1):87–96

Boran FE, Genç S, Kurt M, Akay D (2009) A multi-criteria intuitionistic fuzzy group decision making for supplier selection with TOPSIS method. Expert Syst Appl 36(8):11363–11368

Chang KH, Cheng CH (2010) A risk assessment methodology using intuitionistic fuzzy set in FMEA. Int J Syst Sci 41(12):1457–1471

Chang KH, Cheng CH (2011) Evaluating the risk of failure using the fuzzy OWA and DEMATEL method. J Intell Manuf 22(2):113–129

Chin KS, Wang YM, Poon GKK, Yang JB (2009) Failure mode and effects analysis using a group-based evidential reasoning approach. Comput Oper Res 36(6):1768–1779

Fullér R, Majlender P (2001) An analytic approach for obtaining maximal entropy OWA operator weights. Fuzzy Sets Syst 124(1):53–57

Liu F, Shang YF, Pan LH (2015) A modified TOPSIS method for obtaining the associated weights of the OWA-type operators. Int J Intell Syst 30(10):1101–1116

Liu HC, Liu L, Liu N (2013) Risk evaluation approaches in failure mode and effects analysis: a literature review. Expert Syst Appl 40(2):828–838

Liu HC, Liu L, Li P (2014) Failure mode and effects analysis using intuitionistic fuzzy hybrid weighted Euclidean distance operator. Int J Syst Sci 45(10):2012–2030

Pillay A, Wang J (2003) Modified failure mode and effects analysis using approximate reasoning. Reliab Eng Syst Saf 79(1):69–85

Wang YM, Chin KS, Poon GKK, Yang JB (2009) Risk evaluation in failure mode and effects analysis using fuzzy weighted geometric mean. Expert Syst Appl 36(2):1195–1207
Xu ZS (2005) An overview of methods for determining OWA weights. Int J Intell Syst 20(8): 843–865
Xu ZS (2007a) Intuitionistic fuzzy aggregation operators. IEEE Trans Fuzzy Syst 15(6): 1179–1187
Xu ZS (2007b) Models for multiple attribute decision making with intuitionistic fuzzy information. Int J Uncertainty Fuzziness Knowl Based Syst 15(3):285
Xu ZS (2010) A deviation-based approach to intuitionistic fuzzy multiple attribute group decision making. Group Decis Negot 19(1):57–76
Xu ZS, Yager RR (2006) Some geometric aggregation operators based on intuitionistic fuzzy sets. Int J Gen Syst 35(4):417–433
Yager RR (1988) On ordered weighted averaging aggregation operators in multicriteria decisionmaking. IEEE Trans Syst Man Cybern 18(1):183–190
Zadeh LA (1965) Fuzzy sets. Inf Control 8(3):338–353
Zeng S, Su W (2011) Intuitionistic fuzzy ordered weighted distance operator. Knowl-Based Syst 24(8):1224–1232
Zhang ZF, Chu XN (2011) Risk prioritization in failure mode and effects analysis under uncertainty. Expert Syst Appl 38(1):206–214

Chapter 4
FMEA Using Interval 2-Tuple Hybrid Weighted Distance Measure

The interval 2-tuple linguistic representation method (Zhang 2012) is a useful computational model for computing with words, which has the capability of expressing different types of decision makers' assessment information and has been widely used in many real-world engineering and management problems. Moreover, via this method, decision makers can express their preferences by the use of linguistic term sets with different granularity of uncertainty. Therefore, Liu et al. (2014d) proposed a new risk priority model using interval 2-tuple hybrid weighted distance (ITHWD) measure to overcome the limitations and improve the performance of the traditional FMEA. The new model can not only handle the diversified and uncertain assessments provided by FMEA team members, but also consider the subjective and objective weights of risk factors in the prioritization of the failure modes identified in FMEA. Particularly, the proposed FMEA has exact characteristic and can avoid information distortion and loss in the linguistic information processing.

4.1 Preliminaries

4.1.1 2-Tuple Linguistic Variables

Definition 4.1 Let $S = \{s_0, s_1, \ldots, s_g\}$ be a linguistic term set and $\beta \in [0, 1]$ a value representing the result of a symbolic aggregation operation. Then, the generalized translation function Δ used to obtain the 2-tuple linguistic variable equivalent to β is defined as follows (Tai and Chen 2009; Liu et al. 2014a):

$$\Delta : [0, 1] \rightarrow S \times \left[-\frac{1}{2g}, \frac{1}{2g}\right) \tag{4.1}$$

H.-C. Liu, *FMEA Using Uncertainty Theories and MCDM Methods*,
DOI 10.1007/978-981-10-1466-6_4

$$\Delta(\beta) = (s_i, \alpha), \text{with} \begin{cases} s_i, & i = \text{round}(\beta \cdot g) \\ \alpha = \beta - \frac{i}{g}, & \alpha \in \left[-\frac{1}{2g}, \frac{1}{2g}\right) \end{cases} \tag{4.2}$$

where round(·) is the usual rounding operation, s_i has the closest index label to β and α is the value of the symbolic translation.

Definition 4.2 Let $S = \{s_0, s_1, \ldots, s_g\}$ be a linguistic term set and (s_i, α) be a 2-tuple. There exists a function Δ^{-1}, which is able to convert a 2-tuple linguistic variable into its equivalent numerical value $\beta \in [0, 1]$. The reverse function Δ^{-1} can be defined as follows (Tai and Chen 2009; Liu et al. 2014a):

$$\Delta^{-1} : S \times \left[-\frac{1}{2g}, \frac{1}{2g}\right) \to [0, 1], \tag{4.3}$$

$$\Delta^{-1}(s_i, \alpha) = \frac{i}{g} + \alpha = \beta. \tag{4.4}$$

Clearly, the conversion of a linguistic term into a linguistic 2-tuple consists of adding a value 0 as symbolic translation (Herrera and Martínez 2000):

$$s_i \in S \Rightarrow (s_i, 0). \tag{4.5}$$

Definition 4.3 Let (s_k, α_1) and (s_l, α_2) be two 2-tuples, then (Herrera and Martínez 2000):

(1) If $k < l$ then (s_k, α_1) is smaller than (s_l, α_2);
(2) If $k = l$ then

  (a) if $\alpha_1 = \alpha_2$, then (s_k, α_1) is equal to (s_l, α_2);
  (b) if $\alpha_1 < \alpha_2$ then (s_k, α_1) is smaller than (s_l, α_2);
  (c) if $\alpha_1 > \alpha_2$ then (s_k, α_1) is bigger than (s_l, α_2).

4.1.2 *Interval 2-Tuple Linguistic Variables*

Definition 4.4 Let $S = \{s_0, s_1, \ldots, s_g\}$ be a linguistic term set. An interval 2-tuple linguistic variable is composed of two 2-tuples, denoted by $[(s_i, \alpha_1), (s_j, \alpha_2)]$, where $(s_i, \alpha_1) \leq (s_j, \alpha_2)$. The interval 2-tuple that expresses the equivalent information to an interval value $[\beta_1, \beta_2] (\beta_1, \beta_2 \in [0, 1], \beta_1 \leq \beta_2)$ is derived by the following function (Zhang 2012; Liu et al. 2014c):

$$\Delta[\beta_1,\beta_2]=[(s_i,\alpha_1),(s_j,\alpha_2)] \quad \text{with}\begin{cases} s_i, & i=\text{round}(\beta_1\cdot g)\\ s_j, & j=\text{round}(\beta_2\cdot g)\\ \alpha_1=\beta_1-\frac{i}{g}, & \alpha_1\in\left[-\frac{1}{2g},\frac{1}{2g}\right)\\ \alpha_2=\beta_2-\frac{j}{g}, & \alpha_2\in\left[-\frac{1}{2g},\frac{1}{2g}\right). \end{cases} \tag{4.6}$$

On the contrary, there is always a function Δ^{-1} such that an interval 2-tuple can be converted into an interval value $[\beta_1,\beta_2](\beta_1,\beta_2\in[0,1],\beta_1\le\beta_2)$ as follows:

$$\Delta^{-1}[(s_i,\alpha_1),(s_j,\alpha_2)]=\left[\frac{i}{g}+\alpha_1,\frac{j}{g}+\alpha_2\right]=[\beta_1,\beta_2]. \tag{4.7}$$

Specially, if $s_i=s_j$ and $\alpha_1=\alpha_2$, then the interval 2-tuple linguistic variable reduces to a 2-tuple linguistic variable.

Definition 4.5 For any three interval 2-tuples $\tilde{a}=[(r,\alpha),(t,\varepsilon)]\tilde{a}_1=[(r_1,\alpha_1),(t_1,\varepsilon_1)]$ and $\tilde{a}_2=[(r_2,\alpha_2),(t_2,\varepsilon_2)]$, and let $\lambda\in[0,1]$, then their operations are defined as follows (Liu et al. 2014d):

(1) $\tilde{a}_1\oplus\tilde{a}_2=[(r_1,\alpha_1),(t_1,\varepsilon_1)]\oplus[(r_2,\alpha_2),(t_2,\varepsilon_2)]$
$=\Delta\left[\Delta^{-1}(r_1,\alpha_1)+\Delta^{-1}(r_2,\alpha_2),\Delta^{-1}(t_1,\varepsilon_1)+\Delta^{-1}(t_2,\varepsilon_2)\right]$;

(2) $\lambda\tilde{a}=\lambda[(r,\alpha),(t,\varepsilon)]=\Delta\left[\lambda\Delta^{-1}(r,\alpha),\lambda\Delta^{-1}(t,\varepsilon)\right]$.

Definition 4.6 Let $\tilde{a}_i=[(r_i,\alpha_i),(t_i,\varepsilon_i)](i=1,2,\ldots,n)$ be a set of interval 2-tuples and $w=(w_1,w_2,\ldots,w_n)^T$ be their associated weights, with $w_i\in[0,1],\sum_{i=1}^n w_i=1$. The interval 2-tuple weighted average (ITWA) operator is defined as (Zhang 2012):

$$\begin{aligned}\text{ITWA}_w(\tilde{a}_1,\tilde{a}_2,\ldots,\tilde{a}_n)&=\bigoplus_{i=1}^{n}(w_i\tilde{a}_i)\\ &=\Delta\left[\sum_{i=1}^{n}w_i\Delta^{-1}(r_i,\alpha_i),\sum_{i=1}^{n}w_i\Delta^{-1}(t_i,\varepsilon_i)\right].\end{aligned} \tag{4.8}$$

Definition 4.7 Let $\tilde{a}_i=[(r_i,\alpha_i),(t_i,\varepsilon_i)](i=1,2,\ldots,n)$ be a set of interval 2-tuples, then the interval 2-tuple arithmetic mean is computed as (Liu et al. 2014d):

$$\tilde{\mu}=\Delta\left[\frac{1}{n}\sum_{i=1}^{n}\Delta^{-1}(r_i,\alpha_i),\frac{1}{n}\sum_{i=1}^{n}\Delta^{-1}(t_i,\varepsilon_i)\right]. \tag{4.9}$$

Definition 4.8 Let $\tilde{a}=[(r,\alpha),(t,\varepsilon)]$ and $\tilde{b}=[(r'\alpha'),(t',\varepsilon')]$ be two interval 2-tuples, then

$$d_{\text{ITD}}(\tilde{a},\tilde{b}) = \Delta\sqrt{\frac{1}{2}\left(\left(\Delta^{-1}(r,\alpha) - \Delta^{-1}(r',\alpha')\right)^2 + \left(\Delta^{-1}(t,\varepsilon) - \Delta^{-1}(t',\varepsilon')\right)^2\right)} \tag{4.10}$$

is called the interval 2-tuple distance (ITD) between $\tilde{a}$ and $\tilde{b}$ (Liu et al. 2014d).

Definition 4.9 Let $\tilde{a}_1, \tilde{a}_2, \ldots, \tilde{a}_n$ be a set of interval 2-tuples, and let $\tilde{\mu}$ be the mean of these interval 2-tuples, then we call

$$sim\left(\tilde{a}_{\sigma(j)}, \tilde{\mu}\right) = 1 - \frac{\Delta^{-1} d\left(\tilde{a}_{\sigma(j)}, \tilde{\mu}\right)}{\sum_{i=1}^{n} \Delta^{-1} d\left(\tilde{a}_{\sigma(j)}, \tilde{\mu}\right)}, \quad i = 1, 2, \ldots, n \tag{4.11}$$

the degree of similarity between the jth largest interval 2-tuple $\tilde{a}_{\sigma(j)}$ and the mean $\tilde{\mu}$ (Liu et al. 2014d), where $(\sigma(1), \sigma(2), \ldots, \sigma(n))$ is a permutation of (1, 2, …, n), such that $\tilde{a}_{\sigma(j-1)} \geq \tilde{a}_{\sigma(j)}$ for all j = 2, …, n.

4.2 Interval 2-Tuple Linguistic Hybrid Weighted Distance Measure

Let $\tilde{S}$ be the set of all interval 2-tuples, $\hat{S}$ be the set of all 2-tuples, and $\tilde{A} = \{\tilde{a}_1, \tilde{a}_2, \ldots, \tilde{a}_n\}$ and $\tilde{B} = \{\tilde{b}_1, \tilde{b}_2, \ldots, \tilde{b}_n\}$ be two sets of interval 2-tuples, then the interval 2-tuple weighted distance (ITWD) measure and the interval 2-tuple ordered weighted distance (ITOWD) measure are defined as follows.

Definition 4.10 An ITWD measure of dimension n is a mapping ITWD:$\tilde{S}^n \times \tilde{S}^n \rightarrow \hat{S}$, that has an associated weight vector $w = (w_1, w_2, \ldots, w_n)^T$, with $w_i \in [0, 1]$ and $\Sigma_{i=1}^{n} w_i = 1$, according to the following formula (Liu et al. 2014d):

$$\text{ITWD}\left(\tilde{A}, \tilde{B}\right) = \left(\sum_{i=1}^{n} w_i d_{\text{ITD}}^{\lambda}\left(\tilde{a}_i, \tilde{b}_i\right)\right)^{1/\lambda}, \tag{4.12}$$

where $d_{\text{ITD}}\left(\tilde{a}_i, \tilde{b}_i\right)$ is the interval 2-tuple distance between $\tilde{a}_i$ and $\tilde{b}_i$, and λ is a parameter such that $\lambda \in (-\infty, +\infty) - \{0\}$. Specially, if $\lambda = 1$, then the ITWD measure is reduced to the interval 2-tuple weighted Hamming distance (ITWHD), and if $\lambda = 2$, then the ITWD measure is reduced to the interval 2-tuple weighted Euclidean distance (ITWED).

Definition 4.11 An ITOWD measure of dimension n is a mapping ITOWD: $\tilde{S}^n \times \tilde{S}^n \rightarrow \hat{S}$ that has an associated weighting vector $\omega = (\omega_1, \omega_2, \ldots, \omega_n)^T$, with $\omega_j \in [0, 1]$ and $\Sigma_{j=1}^{n} \omega_j = 1$, according to the following formula (Liu et al. 2014d):

$$\text{ITOWD}(\tilde{A}, \tilde{B}) = \left(\sum_{j=1}^{n} \omega_j d_{\text{ITD}}^{\lambda} \left(\tilde{a}_{\sigma(j)}, \tilde{b}_{\sigma(j)} \right) \right)^{1/\lambda}, \tag{4.13}$$

where $d_{\text{ITD}}(\tilde{a}_{\sigma(j)}, \tilde{b}_{\sigma(j)})$ is the jth largest of the interval 2-tuple distance $d_{\text{ITD}}(\tilde{a}_i, \tilde{b}_i)$, and λ is a parameter such that $\lambda \in (-\infty, +\infty) - \{0\}$. Specially, if $\lambda = 1$, then we obtain the interval 2-tuple ordered weighted Hamming distance (ITOWHD), and if $\lambda = 2$, then we obtain the interval 2-tuple ordered weighted Euclidean distance (ITOWED).

In addition, if there is a tie between $d_{\text{ITD}}(\tilde{a}_i, \tilde{b}_i)$ and $d_{\text{ITD}}(\tilde{a}_j, \tilde{b}_j)$, then we replace each of $d_{\text{ITD}}(\tilde{a}_i, \tilde{b}_i)$ and $d_{\text{ITD}}(\tilde{a}_j, \tilde{b}_j)$ by their average $(d_{\text{ITD}}(\tilde{a}_i, \tilde{b}_i) + d_{\text{ITD}}(\tilde{a}_j, \tilde{b}_j))/2$ in the process of aggregation. If k items are tied, then we replace these by k replicas of their average. The ITOWD is commutative, monotonic, idempotent, and bounded but it does not accomplish always the triangle inequality (Liu et al. 2014d).

By combining the advantages of both the ITWD and the ITOWD measures in the following, Liu et al. (2014d) developed an interval 2-tuple hybrid weighted distance (ITHWD) measure that weights both the given interval 2-tuple distances and their ordered positions.

Definition 4.12 An ITHWD measure of dimension n is a mapping ITWHD: $\tilde{S}^n \times \tilde{S}^n \to \hat{S}$, that has an associated weight vector $\omega = (\omega_1, \omega_2, \ldots, \omega_n)^T$, with $\omega_j \in [0, 1]$ and $\Sigma_{j=1}^{n} \omega_j = 1$, such that

$$\text{ITHWD}(\tilde{A}, \tilde{B}) = \left(\sum_{j=1}^{n} \omega_j d_{\text{ITD}}^{\lambda} \left(\dot{\tilde{a}}_{\sigma(j)}, \dot{\tilde{b}}_{\sigma(j)} \right) \right)^{1/\lambda}, \tag{4.14}$$

where $d_{\text{ITD}}\left(\dot{\tilde{a}}_{\sigma(j)}, \dot{\tilde{b}}_{\sigma(j)}\right)$ is the jth largest of the weighted interval 2-tuple distance $d_{\text{ITD}}(\dot{\tilde{a}}_i, \dot{\tilde{b}}_i)\left(d_{\text{ITD}}\left(\dot{\tilde{a}}_i, \dot{\tilde{b}}_i\right) = n w_i d_{\text{ITD}}(\tilde{a}_i, \tilde{b}_i), i = 1, 2, \ldots, n\right), w = (w_1, w_2, \ldots, w_n)^T$ is the weight vector of $d_{\text{ITD}}(\tilde{a}_i, \tilde{b}_i)(i = 1, 2, \ldots, n)$, with $w_i \in [0, 1]$ and $\sum_{i=1}^{n} w_i = 1$, n is the balancing coefficient, and λ is a parameter such that $\lambda \in (-\infty, +\infty) - \{0\}$. In particular, if $w = (1/n, 1/n, \ldots, 1/n)^T$, then the ITHWD becomes the ITOWD measure and if $\omega = (1/n, 1/n, \ldots, 1/n)^T$, then it becomes the ITWD measure. When $\lambda = \infty$, we obtain the maximum interval 2-tuple weighted distance; when $\lambda = -\infty$, we obtain the minimum interval 2-tuple weighted distance.

Another important issue is determining the weighting vectors associated with the ITHWD measure. In the literature, a lot of methods have been suggested for the determination of the OWA weights (see Sect. 3.1.3), which can also be implemented for the ITHWD measure. In Liu et al. (2014d), the authors defined the ITHWD weights as

$$\omega_j = \frac{sim\left(\tilde{a}_{\sigma(j)}, \tilde{\mu}\right)}{\sum_{j=1}^{n} sim\left(\tilde{a}_{\sigma(j)}, \tilde{\mu}\right)}, \quad j = 1, 2, \ldots, n, \tag{4.15}$$

from which we get $\omega_j \geq 0, j = 1, 2, \ldots, n$, and $\sum_{j=1}^{1} \omega_j = 1$. Note that the weights derived from Eq. (4.15) only depend on the aggregated interval 2-tuple linguistic variables and can relieve the influence of unfair arguments on the aggregated results by assigning low weights to those "false" and "biased" ones, and thus make the aggregated results more reasonable in practical applications.

4.3 The Proposed FMEA Model

In this section, a subjective and objective integrated interval 2-tuple linguistic FMEA model is proposed for the determination of risk priorities of failure modes. The flow diagram of the proposed model is shown in Fig. 4.1.

For a risk analysis problem, suppose there are l decision makers $\mathrm{DM}_k (k = 1, 2, \ldots, l)$ in a FMEA team responsible for the assessment of m failure modes $\mathrm{FM}_i (i = 1, 2, \ldots, m)$ with respect to n risk factors $\mathrm{RF}_j (j = 1, 2, \ldots, n)$. Each decision maker DM_k is given a weight $\lambda_k > 0 (k = 1, 2, \ldots, l)$ satisfying $\sum_{k=1}^{l} \lambda_k = 1$ to reflect his/her relative importance in the risk assessment process. Let $E_k = \left(e_{ij}^k\right)_{m \times n}$ be the linguistic assessment matrix of the kth decision maker, where e_{ij}^k is the linguistic term provided by DM_k on the assessment of FM_i with respect to RF_j. Let w_j^k is the linguistic weight of risk factor RF_j given by DM_k to reflect its relative importance in the prioritization of the failure modes. In addition, FMEA team members may use different linguistic term sets to express their own judgments. Based upon these assumptions and notations, the procedure for the proposed FMEA model is summarized as follows (Liu et al. 2014d):

Step 1. Convert the linguistic assessment matrix $E_k = \left(e_{ij}^k\right)_{m \times n}$ into interval 2-tuple assessment matrix $\tilde{R}_k = \left(\tilde{r}_{ij}^k\right)_{m \times n} = \left(\left[\left(r_{ij}^k, 0\right), \left(t_{ij}^k, 0\right)\right]\right)_{m \times n}$, where $r_{ij}^k, t_{ij}^k \in S, S = \{s_0, s_1, \ldots, s_g\}$ and $r_{ij}^k \leq t_{ij}^k$.

The linguistic information provided in the linguistic assessment matrix E_k can be converted into corresponding interval 2-tuple linguistic assessments according to the transformation method introduced in (Liu et al. 2015).

Step 2. Aggregate the team members' opinions to construct a collective interval 2-tuple assessment matrix $\tilde{R} = \left(\tilde{r}_{ij}\right)_{m \times n}$ and get the aggregated 2-tuple weight vector of risk factors $w = \left[\left(w_j, \alpha_{wj}\right)\right]_{1 \times n}$, where

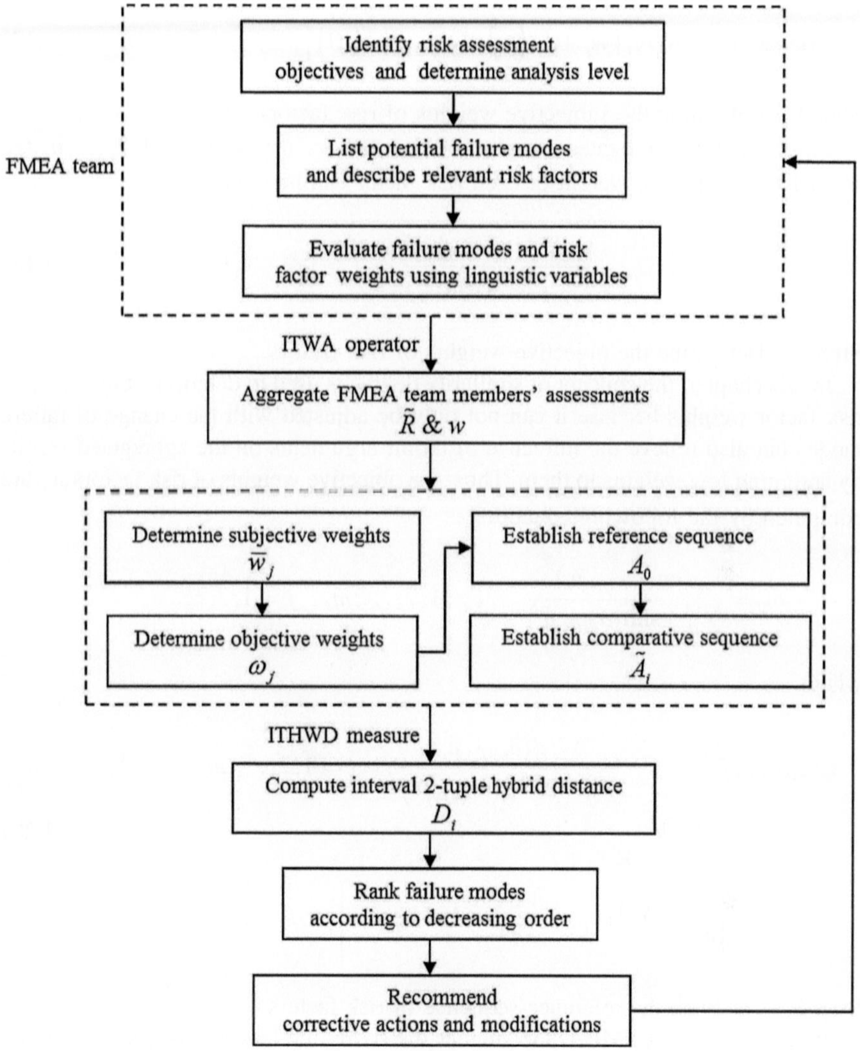

Fig. 4.1 Flow diagram of the proposed FMEA model (Liu et al. 2014d)

$$\begin{aligned}\tilde{r}_{ij} &= \left[(r_{ij}, \alpha_{ij}), (t_{ij}, \varepsilon_{ij})\right] \\ &= \mathrm{ITWA}\left(\left[\left(r_{ij}^{1}, 0\right), \left(t_{ij}^{1}, 0\right)\right], \left[\left(r_{ij}^{2}, 0\right), \left(t_{ij}^{2}, 0\right)\right], \ldots, \left[\left(r_{ij}^{l}, 0\right), \left(t_{ij}^{l}, 0\right)\right]\right) \\ &\quad i = 1, 2, \ldots, m, \quad j = 1, 2, \ldots, n.\end{aligned} \tag{4.16}$$

$$\left(w_j, \alpha_{wj}\right) = \mathrm{TWA}\left[\left(w_j^1, 0\right), \left(w_j^2, 0\right), \ldots, \left(w_j^l, 0\right)\right], \quad j = 1, 2, \ldots, n. \tag{4.17}$$

Step 3. Determine the subjective weights of risk factors

Based on the aggregated weights of risk factors $\left(w_j, \alpha_{wj}\right), j = 1, 2, \ldots, n$, the normalized subjective weight of each risk factor can be obtained as:

$$\bar{w}_j = \frac{\Delta^{-1}\left(w_j, \alpha_{wj}\right)}{\sum_{j=1}^{n} \Delta^{-1}\left(w_j, \alpha_{wj}\right)}, \quad j = 1, 2, \ldots, n. \tag{4.18}$$

Step 4. Determine the objective weights of risk factors

In this chapter, the concept of similarity degree is used to determine the objective risk factor weights because it can not only be adjusted with the change of failure modes but also relieve the influence of unfair arguments on the aggregated results by assigning low weights to them. Thus, the objective weights of risk factors w_j^o are computed by the following equation:

$$\omega_{ij} = \frac{\mathrm{sim}\left(\tilde{a}_{\sigma(ij)}, \tilde{\mu}_i\right)}{\sum_{j=1}^{n} \mathrm{sim}\left(\tilde{a}_{\sigma(j)}, \tilde{\mu}_i\right)}, \quad i = 1, 2, \ldots, m, \quad j = 1, 2, \ldots, n, \tag{4.19}$$

where

$$\mathrm{sim}\left(\tilde{a}_{\sigma(ij)}, \tilde{\mu}_i\right) = 1 - \frac{\Delta^{-1} d\left(\tilde{a}_{\sigma(ij)}, \tilde{\mu}_i\right)}{\sum_{j=1}^{n} \Delta^{-1} d\left(\tilde{a}_{\sigma(ij)}, \tilde{\mu}_i\right)}, \quad i = 1, 2, \ldots, m, \quad j = 1, 2, \ldots, n \tag{4.20}$$

$$\tilde{\mu}_i = \Delta\left[\frac{1}{n}\sum_{j=1}^{n} \Delta^{-1}\left(r_{ij}, \alpha_{ij}\right), \frac{1}{n}\sum_{j=1}^{n} \Delta^{-1}\left(t_{ij}, \varepsilon_{ij}\right)\right], \quad i = 1, 2, \ldots, m. \tag{4.21}$$

Step 5. Establish the reference sequence of risk factors

When conducting FMEA, the smaller the score, the less the risk; therefore, the reference sequence should be the lowest level of the linguistic terms describing the risk factors (Liu et al. 2011, 2014b). In the interval 2-tuple linguistic environment, the minimum 2-tuple $(s_0, 0)$ can be used as the reference value of each risk factor. Thus, the reference sequence is set as:

$$A_0 = \left(r_{0j}\right)_{1 \times n} = \left[(s_0, 0), (s_0, 0), \ldots, (s_0, 0)\right]. \tag{4.22}$$

Step 6. Compute the distances between comparative sequences and the reference sequence

After constructing the collective interval 2-tuple assessment matrix $\tilde{R} = \left(\tilde{r}_{ij}\right)_{m \times n}$, the comparative series with n components or risk factors can be expressed as

$\tilde{A}_i = [\tilde{r}_{i1}, \tilde{r}_{i2}, \ldots, \tilde{r}_{in}], i = 1, 2, \ldots, m$. Then, the distance between the comparative and reference sequences, D_i, can be calculated using the ITHWD measure for each failure mode.

$$D_i = \text{ITHWD}(\tilde{A}_i, A_0) = \left(\sum_{j=1}^{n} \omega_{ij} d_{\text{ITD}}^{\lambda} (\dot{\tilde{r}}_{\sigma(ij)}, \dot{r}_{\sigma(0j)}) \right)^{1/\lambda}, \quad (4.23)$$

where $d_{\text{ITD}}(\dot{\tilde{r}}_{\sigma(ij)}, \dot{r}_{\sigma(0j)})$ is the jth largest of the weighted interval 2-tuple distance $d_{\text{ITD}}(\dot{\tilde{r}}_{ij}, \dot{r}_{0j})$ $(d_{\text{ITD}}(\dot{\tilde{r}}_{ij}, \dot{r}_{0j}) = n\bar{w}_j d_{\text{ITD}}(\tilde{r}_{ij}, r_{0j}), j = 1, 2, \ldots, n)$, n is the balancing coefficient, and λ is a parameter such that $\lambda \in (-\infty, +\infty) - \{0\}$.

Step 7. Determine the ranking order of the failure modes

For FMEA, the bigger the distance obtained from Eq. (4.23), the higher the risk of the failure mode is. As a result, all the failure modes $\text{FM}_i (i = 1, 2, \ldots, m)$ can be prioritized or ranked according to the decreasing order of their ITHWDs $D_i (i = 1, 2, \ldots, m)$.

4.4 An Illustrative Example

4.4.1 Implementation

In this section, a case study of blood transfusion (Lu et al. 2013; Liu et al. 2014d) is presented to illustrate the application of the proposed model for evaluating the risk of healthcare failure modes. Blood transfusion is one of the most routinely performed procedures in hospitals. Although blood transfusion saves lives and reduces morbidities in many clinical diseases and conditions, a significant proportion of adverse events may occur as a result of ordering, collection, transfusion errors, or laboratory errors (Callum et al. 2001). Therefore, identification and prevention of these adverse events are of great importance to optimize the transfusion process and reduce the associated risks. Suppose that the department of blood transfusion in a tertiary care university teaching hospital desires to improve patient care and safety through the use of FMEA, and to prevent and minimize the risk of errors in blood transfusion. Nineteen potential failure modes were initially identified and listed by brainstorming and among them, eleven failure modes $\text{FM}_i (i = 1, 2, \ldots, 11)$ with RPN values greater than 80 were selected for further evaluation. These failure modes, the reasons for them occurring, and their possible effects are presented in Table 4.1. A FMEA team of five medical experts, $\text{DM}_k (k = 1, 2, \ldots, 5)$, has been formed to conduct the risk evaluation and to identify the most serious failure modes for corrective actions. The risk factors, *O*, *S*, and *D*, are considered, which were defined based on historical data and questionnaire answered by all FMEA team members.

Table 4.1 FMEA of the blood transfusion process (Liu et al. 2014d)

No.	Failure mode	Failure cause	Failure effect
1	Insufficient and/or incorrect clinical information on request form	Request form filled out incorrectly/incompletely, patient provided incorrect blood group	Normal process is interrupted; transfusion cannot be performed within appropriate time frame
2	Blood plasma abuse	Blood plasma still used in volume expansion, as nutritional supplement and to improve immunoglobulin levels	Blood resources wasted, risk of transfusion-related reaction and infection increased
3	Insufficient preoperative assessment of the blood product requirement	Improper evaluation of the disease or potential blood loss	Adverse event if compatible blood cannot be prepared in time after emergency cross-matching procedure
4	Blood group verification incomplete	Importance of performing blood group testing on two separate occasions not recognized, use of another sample collected separately or historical records	ABO-incompatible transfusion reaction if no historical blood type or another sample for verification
5	Delivery of blood sample and/or request form delayed	A large number of blood samples have to be delivered to different departments at the same time	Delay in delivery of blood products or reports
6	Incorrect blood components issued	Information or blood product not verified accurately	Blood products cannot be transfused within the appropriate time frame
7	Quality checks not performed on blood products	Insufficient or inaccurate quality checks performed	Poor-quality blood components may be transfused into patients and cause a transfusion reaction
8	Preparation time before infusion >30 min	1. Delivery of blood products to clinic department takes too long: waiting for an elevator, limited staff for delivering blood, blood products are sent to different departments at the same time 2. Infusion is not started in time	Blood components not transfused within 30 min, resulting in reduced quality and associated potential risks to the patient
9	Transfusion cannot be completed within the appropriate time	Transfusion not started when blood products are sent to clinic area; inappropriate transfusion time	Transfusion is delayed and patients receive uncertain quality blood products

(continued)

Table 4.1 (continued)

No.	Failure mode	Failure cause	Failure effect
10	Blood transfusion reaction occurs during the transfusion process	Patient not monitored during the transfusion process	Emergency treatment is delayed, putting the patient's life in danger
11	Bags of blood products are improperly disposed of	Staff unfamiliar with procedures for waste bags	Contamination of environment, traceability cannot be guaranteed if required later

The five team members employ different linguistic term sets to evaluate the risk of failure modes with respect to the three risk factors. Specifically, DM_1 and DM_5 provide their assessments in the set of 5 labels, A; DM_2 and DM_4 provide their assessments in the set of 7 labels, B; DM_3 provides his assessments in the set of 9 labels, C. In addition, the subjective importance of risk factors was rated by the FMEA team with a set of 5 linguistic terms, D. These linguistic term sets are denoted as follows:

$$\begin{aligned}
A = \{ & a_0 = Very\,low\,(VL), a_1 = Low\,(L), a_2 = Moderate\,(M),\ a_3 = High\,(H) \\
& a_4 = Very\,high\,(VH)\}, \\
B = \{ & b_0 = Very\,low\,(VL), b_1 = Low\,(L), b_2 = Moderately\,low\,(ML), b_3 = Moderate\,(M), \\
& b_4 = Moderately\,high(MH), b_5 = High\,(H), b_6 = Very\,high\,(VH)\}. \\
C = \{ & c_0 = Extreme\,low(EL), c_1 = Very\,low(VL), c_2 = Low(L), c_3 = Moderately\,low(ML), \\
& c_4 = Moderate(M), c_5 = Moderately\,high(MH), c_6 = High(H), c_7 = Very\,high(VH), \\
& c_4 = Moderate(M), c_5 = Moderately\,high(MH), c_6 = High(H), c_7 = Very\,high(VH), \\
& c_8 = Extreme\,high(EH)\}, \\
D = \{ & d_0 = Very\,unimportant(VU), d_1 = Unimportant(U), d_2 = Medium(M), \\
& d_3 = Important(I), d_4 = Very\,important(VI)\}.
\end{aligned}$$

The assessments of the eleven failure modes on each risk factor and the importance weights of risk factors provided by the five team members are presented in Tables 4.2 and 4.3, where ignorance information is highlighted and shaded. The five team members are assigned the following relative weights 0.15, 0.20, 0.30, 0.20, and 0.15 in the risk analysis process because of their different domain knowledge and expertise.

Next, we use the proposed FMEA approach to derive the key failure modes in the blood transfusion process. The steps are outlined as follows (Liu et al. 2014d):

Step 1: Transform the linguistic assessment matrix $E_k = \left(e_{ij}^k\right)_{11\times 3}$ into interval 2-tuple assessment matrix $\tilde{R}_k = \left(\tilde{r}_{ij}^k\right)_{11\times 3}$. Taking E_1 as an example, we can get the interval 2-tuple assessment matrix $\tilde{R}_1$ as shown in Table 4.4.

Table 4.2 Linguistic assessments of the eleven failure modes (Liu et al. 2014d)

Decision makes	Risk factors	Failure modes										
		FM1	FM2	FM3	FM4	FM5	FM6	FM7	FM8	FM9	FM10	FM11
DM1	*O*	H	M-H	L-M	L	L	VL	VL	M-H	L-M	L	L-M
	S	H-VH	H	H-VH	VH	M	VH	VH	H-VH	H	H-VH	H
	D	L	M	VH	VL-L	M-H	H-VH	M-H		L-M	H	L
DM2	*O*	MH-H	MH	M-MH	ML-M	L-ML	VL	VL-L	M	ML-M	M	ML
	S	H	M-MH	H	H	M	VH	H	H-VH	H	MH-H	MH-H
	D	L	M	MH-H	L	M-MH	H-VH	M	M	M	H	M
DM3	*O*	MH-H	MH		ML-M	VL-L	VL	EL-L	M-MH	ML	ML	M
	S	VH	M	H-VH	H	M	VH	VH	H-VH	H	VH	H
	D	L	M-MH	H	L	MH	VH	M	M	ML-M	H-VH	ML-M
DM4	*O*	MH	M-MH	M	ML-M	L	VL	VL	MH	ML-M	M	ML-M
	S	MH-H	MH	MH-H	H	M	VH	H	MH-H	MH-H	H	MH
	D	L	M	H	ML	ML	MH-VH	M-MH	M		MH	M-MH
DM5	*O*	M-H	M	M	L-M	L-M	VL	VL	M-H	L	M	L-M
	S	H	M	M-H	H	M	VH	H-VH	H	M-H	H-VH	H
	D	L	M	H	L	M-H	H-VH	M	M	M	H	M

Table 4.3 Linguistic assessments of risk factor weights (Liu et al. 2014d)

Risk factors	Decision makers				
	DM1	DM 2	DM3	DM4	DM5
O	VI	V	V	VI	V
S	VI	VI	VI	VI	VI
D	V	V	V	V	VI

Table 4.4 Interval 2-tuple assessment matrix $\tilde{R}_1$ (Liu et al. 2014d)

Failure modes	Risk factors		
	O	*S*	*D*
FM1	$[(a_3,0), (a_3,0)]$	$[(a_3,0), (a_4,0)]$	$[(a_1,0), (a_1,0)]$
FM2	$[(a_2,0), (a_3,0)]$	$[(a_3,0), (a_3,0)]$	$[(a_2,0), (a_2,0)]$
FM3	$[(a_1,0), (a_2,0)]$	$[(a_3,0), (a_4,0)]$	$[(a_4,0), (a_4,0)]$
FM4	$[(a_1,0), (a_1,0)]$	$[(a_4,0), (a_4,0)]$	$[(a_0,0), (a_1,0)]$
FM5	$[(a_1,0), (a_1,0)]$	$[(a_2,0), (a_2,0)]$	$[(a_2,0), (a_3,0)]$
FM6	$[(a_0,0), (a_0,0)]$	$[(a_4,0), (a_4,0)]$	$[(a_3,0), (a_4,0)]$
FM7	$[(a_0,0), (a_0,0)]$	$[(a_4,0), (a_4,0)]$	$[(a_2,0), (a_3,0)]$
FM8	$[(a_2,0), (a_3,0)]$	$[(a_3,0), (a_4,0)]$	$[(a_0,0), (a_4,0)]$
FM9	$[(a_1,0), (a_2,0)]$	$[(a_3,0), (a_3,0)]$	$[(a_1,0), (a_2,0)]$
FM10	$[(a_1,0), (a_1,0)]$	$[(a_3,0), (a_4,0)]$	$[(a_3,0), (a_3,0)]$
FM11	$[(a_1,0), (a_2,0)]$	$[(a_3,0), (a_3,0)]$	$[(a_1,0), (a_1,0)]$

Table 4.5 2-Tuple linguistic risk factor weights (Liu et al. 2014d)

Risk factors	Decision makers				
	DM1	DM2	DM3	DM4	DM5
O	$(d_4,0)$	$(d_3,0)$	$(d_3,0)$	$(d_4,0)$	$(d_3,0)$
S	$(d_4,0)$	$(d_4,0)$	$(d_4,0)$	$(d_4,0)$	$(d_4,0)$
D	$(d_3,0)$	$(d_3,0)$	$(d_3,0)$	$(d_3,0)$	$(d_4,0)$

Besides, the linguistic evaluations of risk factor weights can be converted into 2-tuple linguistic variables and the results are presented in Table 4.5.

Step 2: The aggregated linguistic ratings of failure modes and the aggregated weights of risk factors are calculated to construct the collective assessment matrix and determine the aggregated weight vector, as in Table 4.6.

Step 3: The subjective weight vector of the three risk factors is computed as $\bar{w} = (0.319, 0.381, 0.300)$ based on Eq. (4.18).

Step 4: The objective weights of risk factors for all the failure modes are computed using Eq. (4.19) as shown in Table 4.7.

Step 5: The reference sequence should be the optimal level of all risk factors for the failure modes in FMEA. Thus, the reference sequence can be determined as $r_0 = (r_O, r_S, r_D) = [\Delta(0), \Delta(0), \Delta(0)]$.

Table 4.6 Collective assessment matrix and aggregated subjective weight vector (Liu et al. 2014d)

Failure modes	*O*	*S*	*D*
FM1	Δ[0.642, 0.750]	Δ[0.788, 0.858]	Δ[0.217, 0.217]
FM2	Δ[0.604, 0.675]	Δ[0.571, 0.604]	Δ[0.500, 0.538]
FM3	Δ[0.313, 0.683]	Δ[0.713, 0.858]	Δ[0.788, 0.821]
FM4	Δ[0.321, 0.463]	Δ[0.821, 0.821]	Δ[0.213, 0.250]
FM5	Δ[0.179, 0.288]	Δ[0.500, 0.500]	Δ[0.504, 0.613]
FM6	Δ[0.038, 0.038]	Δ[0.963, 0.963]	Δ[0.788, 0.963]
FM7	Δ[0.000, 0.071]	Δ[0.858, 0.896]	Δ[0.500, 0.571]
FM8	Δ[0.533, 0.646]	Δ[0.750, 0.892]	Δ[0.425, 0.575]
FM9	Δ[0.321, 0.425]	Δ[0.713, 0.783]	Δ[0.325, 0.600]
FM10	Δ[0.425, 0.425]	Δ[0.788, 0.896]	Δ[0.750, 0.788]
FM11	Δ[0.358, 0.467]	Δ[0.717, 0.750]	Δ[0.425, 0.496]
Weights	Δ(0.838)	Δ(1.000)	Δ(0.788)

Table 4.7 Objective weight vectors for the eleven failure modes (Liu et al. 2014d)

Failure modes	ω_{i1}	ω_{i2}	ω_{i3}
FM1	0.332	0.418	0.250
FM2	0.276	0.466	0.258
FM3	0.351	0.397	0.252
FM4	0.253	0.429	0.319
FM5	0.341	0.403	0.256
FM6	0.357	0.392	0.252
FM7	0.280	0.470	0.250
FM8	0.251	0.434	0.315
FM9	0.262	0.404	0.334
FM10	0.338	0.411	0.251
FM11	0.250	0.406	0.344

Step 6: The distances between the comparative sequences and the reference sequence for the eleven failure modes $D_i, i = 1, 2, \ldots, 11$, are calculated by Eq. (4.23) and let $\lambda = 1$, the results are shown in Table 4.8. In addition, using Eqs. (4.1) and (4.2), we can express the final results in the initial expression domain used by each expert. Taking DM_2 as an example, the final results can be expressed by 2-tuples derived from the linguistic term set *B* with 7 labels, as listed in Table 4.8.

Step 7: Rank all the failure modes in accordance with their ITHWDs in decreasing order. This entails that the failure mode with the largest distance gets the highest priority for attention. The priority ranking of all the failure modes is shown in the last column of Table 4.8.

Table 4.8 ITDs and ITHWDs of failure modes and risk ranking (Liu et al. 2014d)

Failure modes	O	S	D	ITHWD	2-Tuple	Ranking
FM1	$\Delta(0.698)$	$\Delta(0.824)$	$\Delta(0.217)$	$\Delta(0.698)$	$(b_4, 0.031)$	4
FM2	$\Delta(0.641)$	$\Delta(0.588)$	$\Delta(0.519)$	$\Delta(0.669)$	$(b_4, 0.002)$	6
FM3	$\Delta(0.531)$	$\Delta(0.789)$	$\Delta(0.804)$	$\Delta(0.801)$	$(b_5, -0.032)$	1
FM4	$\Delta(0.398)$	$\Delta(0.821)$	$\Delta(0.232)$	$\Delta(0.520)$	$(b_3, 0.020)$	10
FM5	$\Delta(0.240)$	$\Delta(0.500)$	$\Delta(0.561)$	$\Delta(0.473)$	$(b_3, -0.027)$	11
FM6	$\Delta(0.038)$	$\Delta(0.963)$	$\Delta(0.879)$	$\Delta(0.712)$	$(b_4, 0.045)$	3
FM7	$\Delta(0.050)$	$\Delta(0.877)$	$\Delta(0.537)$	$\Delta(0.565)$	$(b_3, 0.065)$	9
FM8	$\Delta(0.592)$	$\Delta(0.824)$	$\Delta(0.506)$	$\Delta(0.687)$	$(b_4, 0.020)$	5
FM9	$\Delta(0.377)$	$\Delta(0.749)$	$\Delta(0.483)$	$\Delta(0.587)$	$(b_4, -0.080)$	7
FM10	$\Delta(0.425)$	$\Delta(0.843)$	$\Delta(0.769)$	$\Delta(0.768)$	$(b_5, -0.065)$	2
FM11	$\Delta(0.416)$	$\Delta(0.734)$	$\Delta(0.462)$	$\Delta(0.566)$	$(b_3, 0.066)$	8

As shown in Table 4.8, the risk ranking of the eleven failure modes is

$$\mathrm{FM}_3 \succ \mathrm{FM}_{10} \succ \mathrm{FM}_6 \succ \mathrm{FM}_1 \succ \mathrm{FM}_8 \succ \mathrm{FM}_2 \succ \mathrm{FM}_9 \succ \mathrm{FM}_{11} \succ \mathrm{FM}_7 \succ \mathrm{FM}_4 \succ \mathrm{FM}_5.$$

Hence, FM3 is the most critical failure mode and should be given the top priority for correction by the hospital; this will be followed by FM10, FM6, FM1, FM8, FM2, FM9, FM11, FM7, FM4, and FM5.

4.4.2 Sensitivity Analysis

In the above analysis, we set the parameter $\lambda = 1$ in the application of the ITHWD measure. In this section, a sensitivity analysis by changing the parameter λ is calculated according to the information given in Tables 4.6 and 4.7. Depending on the steps of the proposed FMEA model with different λ values, we can obtain the ITHWD for each failure mode together with their equivalent numerical values as shown in Fig. 4.2, where λ is set to $[-10, 10]$. From Fig. 4.2, it can be observed that the values of ITHWDs are non-decreasing with respect to λ. In addition, the risk priority rankings of failure modes may be different with the change of the parameter λ. For most of the cases, the most serious failure mode is FM3 because it seems to be the one with the biggest distance to the reference sequence. However, for some particular cases, we may find another high-risk failure mode. For example, with $\lambda = 3$, FM6 becomes the most critical failure because it has the highest rating of severity.

Since the variation of λ value may lead to different ranking orders of failure modes, a decision maker may have difficulty in identifying the most important failures with different parameter values λ. In other words, it is necessary for the

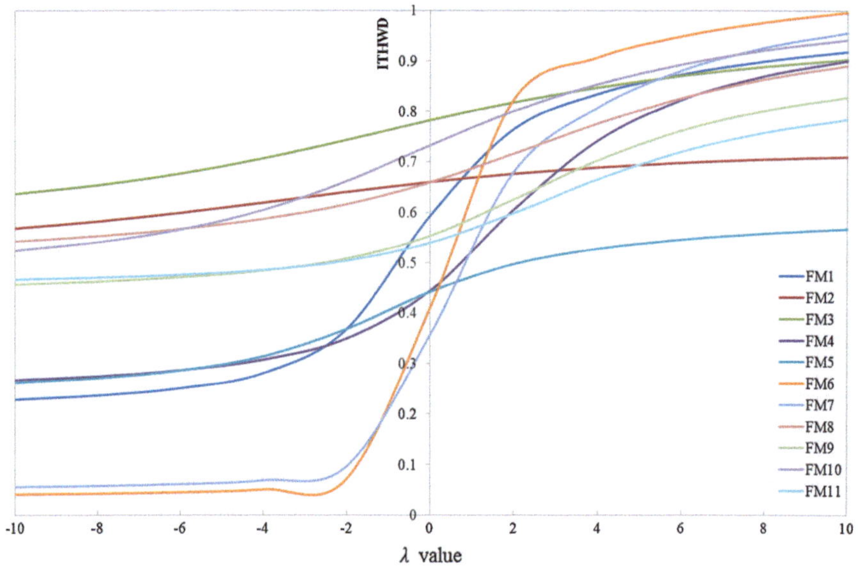

Fig. 4.2 Interval 2-tuple hybrid weighted distances with $\lambda \in [-10, 10]$ (Liu et al. 2014c)

decision maker to set λ value before information aggregation. In general, the more pessimistic of the decision maker, the larger λ value he or she may set, which means each failure mode is associated with a higher evaluation value among the risk factors. On the contrary, the more optimistic of the decision maker, the smaller λ value he or she may set. If the decision maker cannot give his or her subjective preference, then the most commonly used value $\lambda = 1$ can be taken. Therefore, by using the ITHWD measure, the attitudinal character of the decision maker can be taken into account when conducting the FMEA process.

4.4.3 Comparison and Discussion

For presenting the strong points of the proposed FMEA approach, a comparison of the results with the conventional RPN method and the fuzzy VIKOR (Liu et al. 2012) is made in this part. Table 4.9 exhibits the ranking results of all the eleven failure modes as obtained using these three approaches.

From Table 4.9, we can see that there is a great difference between the two sets of risk priority rankings produced by the conventional RPN method and the proposed FMEA model. Except for FM3, the rank orders of the rest failure modes obtained by the proposed approach are all different from those by the traditional FMEA. This can be explained by the shortcomings of the conventional RPN method, which lead to biased or even misleading conclusions. For example, both FM6 and FM7 have the same RPN = 80. Namely, the failure modes with different

Table 4.9 Ranking comparison (Liu et al. 2014d)

Failure modes	O	S	D	RPN	Ranking RPN	Fuzzy VIKOR	Ranking ITHWD
FM1	6	7	3	126	5	4	4
FM2	6	6	5	180	4	7	6
FM3	5	7	7	245	1	2	1
FM4	5	7	3	105	8	8	10
FM5	3	5	6	90	9	11	11
FM6	1	10	8	80	10	1	3
FM7	2	8	5	80	10	6	9
FM8	6	8	5	240	2	5	5
FM9	4	7	4	112	6	10	7
FM10	4	8	7	224	3	3	2
FM11	4	7	4	112	6	9	8

combinations of *O*, *S*, and *D* produce the identical RPN value, leading to difficult decision making by the traditional FMEA for the priority of corrective actions. However, this problem can be easily solved by applying the proposed interval 2-tuple FMEA model. According to their ITHWDs, for FM6, more urgently corrective (or preventive) actions are needed.

The effect of risk factor weights introduced in the proposed model can be clearly seen in the results obtained for FM1 and FM2, where *O*, *S*, and *D* are assigned 6, 7, 3 and 6, 6, 5, respectively. In this example, we can find that *O* is 6 for both the failure modes, FM1 has a higher value of S, and in FM2 the value of *D* is higher than FM2's. According to the conventional RPN method, FM1 (RPN = 126) is ranked behind FM2 (RPN = 180) and thus given a lower priority. However, in practice, FM1 is more important because it has a higher severity rating and more weight is given to the risk factor of severity in the healthcare context. Using the proposed approach, the ranking of FM1 is 4, and it has a higher priority in comparison with FM2. This shows that a more accurate ranking can be achieved by applying the ITHWD measure to FMEA.

Second, there are some differences between the risk-ranking orders derived by the fuzzy VIKOR method and the proposed risk priority model. These inconsistent ranking results can be understood from the fact that the objective weights of risk factors are not considered during the fuzzy VIKOR-based risk analysis, which may result in unreasonable ranking of failure modes. For example, according to the fuzzy VIKOR, FM9 is ranked behind FM4. In reality, however, the former is more important, and thus, the result of the proposed method suggests that FM9 has a higher priority in comparison with FM4. This is also true for FM2 and FM7. Besides, FM8 turned out to be the most critical failure mode according to the fuzzy VIKOR method, while by using the proposed FMEA, it ranks the third position and FM3 becomes the most important one at the same time. Giving FM3 the top priority can also be validated by the conventional RPN method. In addition, the fuzzy group assessments given by the FMEA team members are defuzzified at the very beginning of the fuzzy VIKOR algorithm. This may lead to loss some information in the following risk analysis process and hence, a lack of precision in the final results.

The example presented above has demonstrated the effectiveness of the proposed approach using interval 2-tuple hybrid weighted distance measure for the prioritization of failure modes in the healthcare environment. Comparing the conventional RPN method and its various improvements such as fuzzy logic-based FMEA, the risk priority model proposed has the following advantages: (1) The proposed FMEA has exact characteristic in linguistic information processing, which can effectively avoid the loss and distortion of information in the processing of linguistic terms. (2) Both subjective and objective weights of risk factors are taken into account in the determination of risk priority of failure modes. (3) By using the ITHWD measure, the attitudinal character of the decision maker can be considered when conducting the risk assessment process. (4) Risk factors and their subjective relative weights are evaluated in a linguistic manner rather than in crisp numbers, and FMEA team members can provide their assessments through multi-granularity linguistic term sets. (5) The fuzzy, uncertain, and incomplete assessment information on risk factors provide by different experts can be well reflected and handled using the interval the 2-tuple linguistic variables. (6) The proposed approach can achieve a more accurate risk priority ranking and discriminate among the results far more accurate, thus providing more effective information to assist the risk decision making.

References

Callum JL, Kaplan HS, Merkley LL, Pinkerton PH, Rabin Fastman B, Romans RA, Coovadia AS, Reis MD (2001) Reporting of near-miss events for transfusion medicine: improving transfusion safety. Transfusion 41(10):1204–1211

Herrera F, Martínez L (2000) A 2-tuple fuzzy linguistic representation model for computing with words. IEEE Trans Fuzzy Syst 8(6):746–752

Liu HC, Liu L, Bian QH, Lin QL, Dong N, Xu PC (2011) Failure mode and effects analysis using fuzzy evidential reasoning approach and grey theory. Expert Syst Appl 38(4):4403–4415

Liu HC, Liu L, Liu N, Mao LX (2012) Risk evaluation in failure mode and effects analysis with extended VIKOR method under fuzzy environment. Expert Syst Appl 39(17):12926–12934

Liu HC, Lin QL, Wu J (2014a) Dependent interval 2-tuple linguistic aggregation operators and their application to multiple attribute group decision making. Int J Uncertainty, Fuzziness Knowl-Based Syst 22(5):717–735

Liu HC, Liu L, Li P (2014b) Failure mode and effects analysis using intuitionistic fuzzy hybrid weighted Euclidean distance operator. Int J Syst Sci 45(10):2012–2030

Liu HC, Ren ML, Wu J, Lin QL (2014c) An interval 2-tuple linguistic MCDM method for robot evaluation and selection. Int J Prod Res 52(10):2867–2880

Liu HC, You JX, You XY (2014d) Evaluating the risk of healthcare failure modes using interval 2-tuple hybrid weighted distance measure. Comput Ind Eng 78:249–258

Liu HC, Li P, You JX, Chen YZ (2015) A novel approach for FMEA: Combination of interval 2-tuple linguistic variables and grey relational analysis. Qual Reliab Eng Int 31(5):761–772

Lu Y, Teng F, Zhou J, Wen A, Bi Y (2013) Failure mode and effect analysis in blood transfusion: a proactive tool to reduce risks. Transfusion 53(12):3080–3087

Tai WS, Chen CT (2009) A new evaluation model for intellectual capital based on computing with linguistic variable. Expert Syst Appl 36(2, Part 2):3483–3488

Zhang H (2012) The multiattribute group decision making method based on aggregation operators with interval-valued 2-tuple linguistic information. Math Comput Model 56(1–2):27–35

Chapter 5
FMEA Using Fuzzy Evidential Reasoning and GRA Method

Two most important issues of FMEA are the acquirement of FMEA team members' diversity assessments and the determination of risk priorities of the identified failure modes. First, FMEA team members often demonstrate different opinions and knowledge and produces different types of assessment information because of their different expertise and backgrounds. Second, the traditional FMEA which determines the risk priorities of failure modes by using RPNs has been criticized to have many shortcomings. Therefore, Liu et al. (2011) presented a new risk priority model for FMEA based on fuzzy evidential reasoning (FER) and grey relation analysis (GRA) method to improve the effectiveness of the traditional FMEA. The proposed FMEA can not only capture FMEA team members' diversity opinions under different types of uncertainties and incorporate the importance weights of risk factors into the prioritization of failure modes, but also take advantage of the benefits of fuzzy logic and grey theory without the need of asking experts too much.

5.1 Fuzzy Evidential Reasoning Approach

The evidential reasoning (ER) approach was developed by combing the Dempster–Shaffer (D–S) theory (Shafer 1976) with a distributed modeling framework for dealing with multi-criteria decision-making (MCDM) problems characterized by both quantitative and qualitative attributes with various types of uncertainties (Yang et al. 2006; Guo et al. 2009). Its main advantage is that both precise data and subjective judgments with uncertainty can be consistently modeled under a unified framework. The ER approach provides a novel procedure for aggregating multiple criteria based on the distributed assessment framework and the evidence combination rule of D–S theory.

Extensive research dedicated to the ER approach has been conducted in recent years. Experiences show that a decision maker may not always be confident enough to provide subjective assessments to individual grades only but at times wishes to

H.-C. Liu, *FMEA Using Uncertainty Theories and MCDM Methods*,
DOI 10.1007/978-981-10-1466-6_5

be able to assess beliefs to subsets of adjacent grades. It is to deal with the problem that the interval-grade ER (IER) approach is proposed (Yang and Singh 1994). Another extension to the original ER approach is to take account of vagueness or fuzzy uncertainty, i.e., the assessment grades are no longer clearly distinctive crisp sets, but are defined as dependent fuzzy sets. Yang et al. (2006) proposed the fuzzy ER approach (FER) to extend the original ER individual grades to fuzzy grades to capture fuzziness caused by the fuzzy evaluation grades. Guo et al. (2009) developed a general ER modeling framework and an attribute aggregation process, which is referred to as the fuzzy IER (FIER) algorithm, to deal with both fuzzy and interval-grade assessments. For the MCDM problem with unknown criteria weights, Fu and Chin (2014) proposed a robust ER approach to compare alternatives by measuring their robustness with respect to criteria weights and generate a robust solution in the ER context. Chen et al. (2016) proposed a new fuzzy MCDM method based on intuitionistic fuzzy sets and ER methodology, in which the ER methodology is used to aggregate each decision maker's decision matrix to get the aggregated decision matrix.

In this chapter, the FER approach is used to deal with the diversity and uncertainty of assessment information given by FMEA members, and the involved steps are presented as follows (Liu et al. 2011):

Step 1. Assess risk factors using belief structures

The three risk factors O, S, and D can be evaluated numerically or linguistically. Both of them have been extensively applied and have their merits and demerits. However, there is a high level of uncertainty involved in FMEA since it is a group decision behavior and the assessment information for risk factors mainly based on experts' subjective judgments may be complete or incomplete, precise or imprecise, and certain or uncertain. In addition, most experts are willing to express their opinions by belief degrees (or possibility measures) based on a set of evaluation grades, i.e., {Very Low, Low, Moderate, High, and Very High}. As such, in this chapter, we choose linguistic terms for the assessment of risk factors and the individual evaluation grade set is defined as a fuzzy set H_F as follows:

$$\begin{aligned} H_F &= \{H_{11}, H_{22}, H_{33}, H_{44}, H_{55}\} \\ &= \{\text{Very Low, Low, Moderate, High, Very High}\}. \end{aligned}$$

In order to generalize the $\hat{H}_F = \left\{H_{pq}, p = 1,\ldots,5; q = 1,\ldots,5\right\}$ to fuzzy sets, we assume that a general set of fuzzy individual assessment grades $\left\{H_{pp}\right\}$, $p = 1,\ldots,5$ are dependent on each other and only two adjacent fuzzy individual assessment grades may intersect. Based on experts' opinions, we can approximate all the five individual assessment grades by trapezoidal fuzzy numbers for simplifying the discussion and without loss of generality, and their membership function values can be determined according to the historical data and the detailed questionnaire answered by all experts, as shown in Fig. 5.1 and Table 5.1.

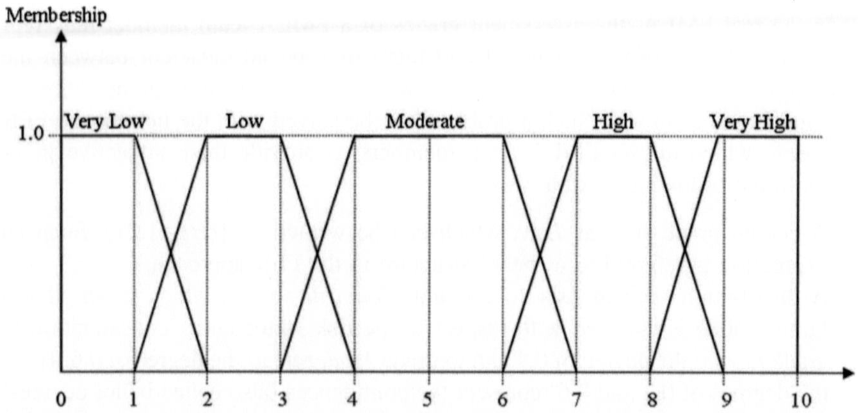

Fig. 5.1 Fuzzy membership function for linguistic terms (Liu et al. 2011)

Table 5.1 Linguistic terms for rating failure modes

Linguistic terms	Fuzzy number
Very low	(0, 0, 1, 2)
Low	(1, 2, 3, 4)
Moderate	(3, 4, 6, 7)
High	(6, 7, 8, 9)
Very high	(8, 9, 10, 10)

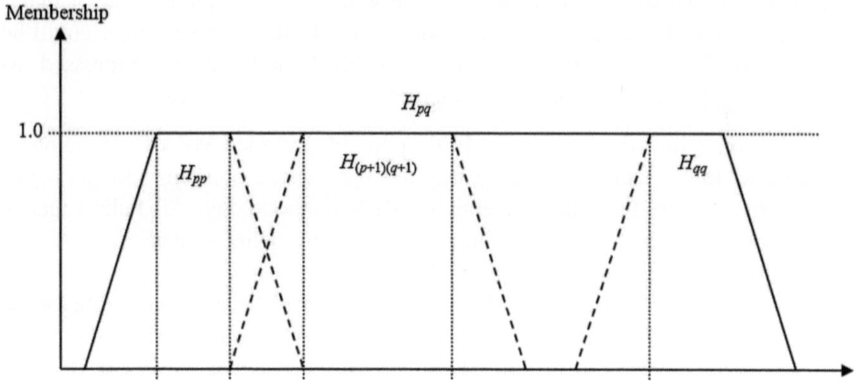

Fig. 5.2 Interval fuzzy grades set (Liu et al. 2011)

Furthermore, we define the interval fuzzy assessment grades sets H_{ij} for $p = 1, \ldots, 4$ and $q = p + 1$ to 5 as trapezoidal fuzzy sets that include fuzzy individual grades H_{pp}, $H_{(p+1)(p+1)}$, …, H_{qq}. If the individual assessment grades are trapezoidal fuzzy sets, every interval grade will be a trapezoidal fuzzy set as shown in Fig. 5.2.

In the real FMEA, the assessment grades of a FMEA team member may represent a vague concept or standard and there may be no clear cut between the meanings of two adjacent grades. In other words, these evaluation grades may not be regarded as crisp sets. Such a problem can be solved with the help of the FER approach, which allows FMEA team members to provide their subjective judgments in the following flexible ways:

- A certain grade such as *Low*, which can be written as $\{(H_{22}, 1.0)\}$. Such an expression is referred to as belief structure in the FER approach.
- A distribution such as *Low* to 0.4 and *Moderate* to 0.6, which means that a failure mode is assessed with respect to the risk factor under consideration to grade *Low* to the degree of 0.4 and to grade *Moderate* to the degree of 0.6. Here, the degrees of 0.4 and 0.6 represent the confidences (also called belief degrees) of the FMEA team member in his/her subjective judgments and the distribution can be equivalently expressed as $\{(H_{22}, 0.4), (H_{33}, 0.6)\}$. When all the confidences are summed to one, the distribution is said to be complete; otherwise, it is said to be incomplete. For example, $\{(H_{22}, 0.4), (H_{33}, 0.5)\}$ is an incomplete distribution or called incomplete assessment, where the missing information of 0.1 is referred to as local ignorance and could be assigned to any grade between *Very Low–Very High* according to the D–S theory (Shafer 1976).
- An interval such as *Low–Moderate*, which means that the grade of a failure mode with respect to the risk factor under evaluation is between *Low* and *Moderate*. This can be written as $\{(H_{23}, 1.0)\}$.
- No judgment, which means the FMEA team member is not willing to or cannot provide an assessment for a failure mode with respect to the risk factor under consideration. In other words, the grade by this FMEA team member could be anywhere between *Very Low* and *Very High* and can be expressed as $\{(H_{15}, 1.0)\}$. Such judgments are referred to as total ignorance.

Obviously, belief structures in the FER approach provide FMEA team members with an easy-to-use and very flexible way to express their opinions and can better quantify risk factors than the conventional RPN methodology. All failure modes with respect to the risk factors can be evaluated using belief structures.

Step 2. Compute the fuzzy group belief assessment matrix

Suppose there are l members (TM_1, …, TM_l) in a FEMA team responsible for the assessment of m failure modes (FM_1, …, FM_m) with respect to n risk factors (RF_1, …, RF_n). Each team member TM_k is given a weight $\lambda_k > 0$ ($k = 1, \ldots, l$) satisfying $\sum_1^l \lambda_k = 1$ to reflect his/her relative importance in the FMEA team. Let $\tilde{w}_j^k = \left(w_{ja}^k, w_{jb}^k, w_{jd}^k\right)$ is the weight of risk factor RF_j given by TM_k to reflect its relative importance in the determination of risk priorities of the failure modes. Since they are not easy to be precisely determined due to the same reason as risk factors, the relative importance weights of risk factors are assessed using the linguistic terms

Table 5.2 Linguistic terms for rating risk factor weights

Linguistic terms	Fuzzy number
Very low (VL)	(0, 0, 0.25)
Low (L)	(0, 0.25, 0.5)
Moderate (M)	(0.25, 0.5, 0.75)
High (H)	(0.5, 0.75, 1)
Very high (VH)	(0.75, 1, 1)

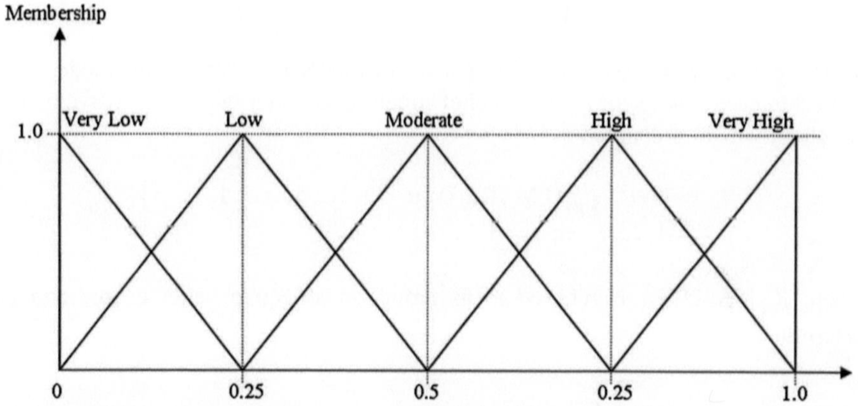

Fig. 5.3 Membership functions of fuzzy weights (Liu et al. 2011)

in Table 5.2, whose membership functions are visualized in Fig. 5.3. The group weight of risk factor RF_j of the l team members is denoted as

$$\tilde{w}_j = \sum_{k=1}^{l} \lambda_k \tilde{w}_j^k = \left(\sum_{k=1}^{l} \lambda_k w_{ja}^k, \sum_{k=1}^{l} \lambda_k w_{jb}^k, \sum_{k=1}^{l} \lambda_k w_{jd}^k \right), \quad j = 1, 2, \ldots, n. \tag{5.1}$$

The group weights of risk factors are first defuzzified using Eq. (5.7) and then normalized by

$$\bar{w}_j = \frac{w_j}{\sum_{j=1}^{n} w_j}, \quad j = 1, 2, \ldots, n, \tag{5.2}$$

where w_j is referred to as the crisp number of the group risk factor weight $\tilde{w}_j$.

Let $\left\{ \left(H_{pq}, \beta_{pq}^k(\mathrm{FM}_i, \mathrm{RF}_j) \right), p = 1, \ldots, 5;\, q = 1, \ldots, 5 \right\}$ be the belief structure provided by TM_k on the assessment of FM_i with respect to RF_j, where H_{pp} for $p = 1, \ldots, 5$ are fuzzy assessment grades defined for risk assessment, H_{pq} for $p = 1, \ldots, 4$ and $q = p + 1$ to 5 are the intervals fuzzy assessment grades between H_{pp} and H_{qq}, and $\beta_{pq}^k(\mathrm{FM}_i, \mathrm{RF}_j)$ are the belief degrees to which FM_i assessed on

RF_j to the intervals H_{pq}. All the grades H_{pp} for $p = 1, \ldots, 5$ and the intervals H_{pq} for $p = 1, \ldots, 4$ and $q = p + 1$ to 5 together form the frame of discernment, which is expressed as $\hat{H}_F = \{H_{pq}, p = 1, \ldots, 5;\ q = 1, \ldots, 5\}$, or equivalently

$$\hat{H}_F = \left\{ \begin{matrix} H_{11} & H_{12} & H_{13} & H_{14} & H_{15} \\ & H_{22} & H_{23} & H_{24} & H_{25} \\ & & H_{33} & H_{34} & H_{35} \\ & & & H_{44} & H_{45} \\ & & & & H_{55} \end{matrix} \right\}. \tag{5.3}$$

The collective assessment of the l team members for each failure mode with respect to each risk factor is also a belief structure, called group or collective belief structure, which is denoted as

$$\begin{gathered} X_{ij} = \left\{ \left(H_{pq}, \beta_{pq}(\mathrm{FM}_i, \mathrm{RF}_j)\right), p = 1, \ldots, 5;\ q = 1, \ldots, 5 \right\}, \\ i = 1, 2, \ldots, m;\ j = 1, 2, \ldots, n, \end{gathered} \tag{5.4}$$

where $\beta_{pq}(\mathrm{FM}_i, \mathrm{RF}_j)$ is referred to as group or collective belief degree and is determined by

$$\begin{gathered} \beta_{pq}(\mathrm{FM}_i, \mathrm{RF}_j) = \sum_{k=1}^{l} \lambda_k \beta_{pq}^k(\mathrm{FM}_i, \mathrm{RF}_j), \\ p = 1, 2, \ldots, 5;\ q = 1, 2, \ldots, 5;\ i = 1, 2, \ldots, m;\ j = 1, 2, \ldots, n. \end{gathered} \tag{5.5}$$

That is, a group belief degree is the weighted sum of the individual belief degrees corresponding to the same grade or interval. In addition, the group belief structures for m failure modes with respect to n risk factors form a fuzzy group belief assessment matrix as shown in Eq. (5.6), which differs from the traditional assessment matrix in that it consists of both fuzzy assessment grades and belief structures.

$$\tilde{X} = \begin{bmatrix} \tilde{X}_{11} & \tilde{X}_{12} & \cdots & \tilde{X}_{1n} \\ \tilde{X}_{21} & \tilde{X}_{22} & \cdots & \tilde{X}_{2n} \\ \vdots & \vdots & \cdots & \vdots \\ \tilde{X}_{m1} & \tilde{X}_{m2} & \cdots & \tilde{X}_{mn} \end{bmatrix} \tag{5.6}$$

Step 3. Obtain the crisp group belief assessment matrix

Based on the fuzzy group belief assessment matrix $\tilde{X}$, group belief structures on the assessment of each failure mode with respect to the n risk factors can be aggregated into an overall belief structure using the defuzzification method and the weighted average method successively. Chen and Klein (1997) have proposed an easy defuzzification method for obtaining the crisp number of a fuzzy set, which is shown here in Eq. (5.7).

$$h_{pq} = \frac{\sum_{r=0}^{g} (b_r - c)}{\sum_{r=0}^{g} (b_r - c) - \sum_{r=0}^{g} (a_r - d)}, \quad p = 1, 2, \ldots, 5;\ q = 1, 2, \ldots, 5, \quad (5.7)$$

where g is the number of α-levels and h_{pq} is the defuzzified crisp number of H_{pq}.

Finally, the overall assessment of the failure mode FM_i with respect to the risk factor RF_j is also a crisp number, called overall belief structure, which can be aggregated by the following equation:

$$X_{ij} = \sum_{p=1}^{5} \sum_{q=1}^{5} h_{pq} \beta_{pq} (\mathrm{FM}_i, \mathrm{RF}_j), \quad i = 1, 2, \ldots, m;\ j = 1, 2, \ldots, n. \quad (5.8)$$

Consequently, the fuzzy group belief assessment matrix $\tilde{X}$ can be defuzzified to get the crisp group belief assessment matrix X, which is shown as follows:

$$X = \begin{bmatrix} X_{11} & X_{12} & \cdots & X_{1n} \\ X_{21} & X_{22} & \cdots & X_{2n} \\ \vdots & \vdots & \cdots & \vdots \\ X_{m1} & X_{m2} & \cdots & X_{mn} \end{bmatrix}. \quad (5.9)$$

5.2 The GRA Method

The grey theory, first proposed by Deng (1989), deals with decisions characterized by incomplete information, such as operation, mechanism, structure, and behavior, which are neither deterministic nor totally unknown, but are partially known. It explores system behavior using relation analysis and model construction. The use of grey relation analysis (GRA) within the FMEA framework is practicable and can be accomplished (Chang et al. 2001; Liu et al. 2013, 2015).

Next, the GRA method is adopted to rank the failure modes identified in FMEA based on the results of the FER approach. The procedure of GRA is expounded as follows (Liu et al. 2011):

Step 1. Generate the comparative series

An information series with n components or risk factors can be expressed as $X_i' = (X_{i1}', X_{i2}', \ldots, X_{in}')$, where X_{in}' denotes the jth risk factor of X_i'. If all information series are comparable, the m information series can be described as the following matrix:

$$X' = \begin{bmatrix} X'_{11} & X'_{12} & \cdots & X'_{1n} \\ X'_{21} & X'_{22} & \cdots & X'_{2n} \\ \vdots & \vdots & \cdots & \vdots \\ X'_{m1} & X'_{m2} & \cdots & X'_{mn} \end{bmatrix}. \tag{5.10}$$

For the application of this matrix in FMEA, the matrix X' is generated based on the crisp group belief assessment matrix X, which is determined by Eq. (5.9).

Step 2. Determine the standard series

Degree of relation can describe the relationship of two series; thus, an objective series called the standard series shall be established and expressed as $X_0 = (X_{01}, X_{02}, \ldots, X_{0n})$. When conducting FMEA, the smaller the score, the less the risk; therefore, the standard series can be the lowest level of all the risk factors:

$$\begin{aligned} X_0 &= (X_{01}, X_{02}, \ldots, X_{0n}) = [H_{11}, H_{11}, \ldots, H_{11}] \\ &= [h_{11}, h_{11}, \ldots, h_{11}] \end{aligned} \tag{5.11}$$

Step 3. Compute the difference between comparative series and standard series

The difference between the comparative and the standard series, D_0, is calculated and reflected in a form of matrix as seen below:

$$D_0 = \begin{bmatrix} \Delta_{11} & \Delta_{12} & \cdots & \Delta_{1n} \\ \Delta_{22} & \Delta_{22} & \cdots & \Delta_{2n} \\ \vdots & \vdots & \cdots & \vdots \\ \Delta_{m1} & \Delta_{m2} & \cdots & \Delta_{mn} \end{bmatrix}, \tag{5.12}$$

where $\Delta_{ij} = \left\| X'_{0j} - X_{ij} \right\|$.

Step 4. Calculate the grey relation coefficient

The grey relation coefficient, γ_{ij}, is calculated using Eq. (5.13) for each risk factor of the failure modes identified in the FMEA.

$$\gamma_{ij} = \frac{\Delta_{\min} + \zeta\Delta_{\max}}{\Delta_{ij} + \zeta\Delta_{\max}}, \quad i = 1, 2, \ldots, m;\ j = 1, 2, \ldots, n, \tag{5.13}$$

where $\Delta_{\min} = \min_i \min_j (\Delta_{ij})$, $\Delta_{\max} = \max_i \max_j (\Delta_{ij})$, and ζ is an identifier, $\zeta \in (0, 1)$, only affecting the relative value of risk without changing the priority. Generally, ζ can be 0.5 (Deng 1989).

Step 5. Determine the degree of relation

This step is to obtain the degree of grey relation based upon the grey relation coefficients γ_{ij} and the group weights of risk factors $\bar{w}_j$, which is determined by Eq. (5.2). The degree of grey relation is calculated for each failure mode using the following formulation

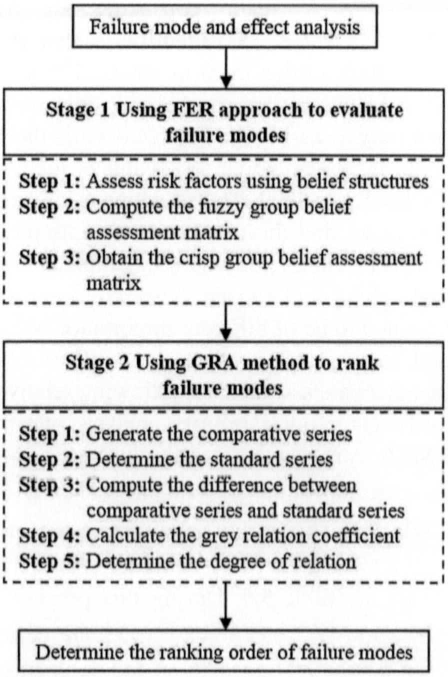

Fig. 5.4 Flowchart of the proposed FMEA model

$$\Gamma_{ij} = \sum_{j=1}^{n} \bar{w}_j \gamma_{ij} \tag{5.14}$$

The degree of relation in FMEA represents the relationship between potential failure modes and the optimal value of risk factors. The higher the degree of relation obtained from Eq. (5.14), the smaller the effect of the failure mode. As a result, all the failure modes can be ranked according to the degree of grey relation of each failure mode.

To sum up, the FMEA model proposed by Liu et al. (2011) based on the FER and the GRA methods can be delineated using the flowchart in Fig. 5.4.

5.3 An Illustrative Example

In this section, we provide a numerical example to illustrate the potential applications of the proposed FMEA and particularly the potentials of using the FER and the GRA method in capturing FMEA team members' diversity opinions and prioritizing failure modes under different types of uncertainties. The FMEA example is adapted from Wang et al. (2009), Liu et al. (2011).

A FMEA team consisting of five cross-functional team members, TM_k $(k = 1, 2, \ldots, 5)$, identifies seven potential failure modes in a system and needs to prioritize them in terms of risk factors such as *O*, *S*, and *D* so that high risky failure modes can be corrected with top priorities. Due to the difficulty in precisely assessing the risk factors and their relative importance weights, the FMEA team members agree to evaluate them using the linguistic terms defined in Tables 5.1 and 5.2. The assessment information of the seven failure modes on each risk factor and the risk factor weights provided by the five team members is presented in Table 5.3, where incomplete assessments and ignorance information are highlighted and shaded. The five team members from different departments are assumed to be of different importance because of their different domain knowledge and expertise. To reflect their differences in performing FEMA, the five team members are assigned the following relative weights: 0.15, 0.20, 30, 0.25, and 0.10.

To carry out a priority analysis, we first use belief structures to express the FMEA team members' individual assessments and synthesize them to construct the fuzzy group belief assessment matrix $\tilde{X} = \left[\tilde{X}_{ij}\right]_{7\times3}$ by Eq. (5.6), as presented in Table 5.4. The group belief structures in the matrix $\tilde{X}$ are then defuzzified and aggregated into overall belief structures using Eqs. (5.7) and (5.8). The results are shown in Table 5.5. During this process, all the fuzzy assessment grades $\hat{H}_F = \left\{H_{pq}, p = 1, \ldots, 5;\ q = 1, \ldots, 5\right\}$ are defuzzified by using Eq. (5.7) to produce a crisp number. The results of the defuzzification are tabulated in Table 5.6.

Next, the data in Table 5.5 are analyzed using the GRA method. The comparative series is generated based on the table using Eqs. (5.9) and (5.10), as seen in the matrix below

$$X' = X = \begin{bmatrix} 0.379 & 0.364 & 0.210 \\ 0.541 & 0.438 & 0.170 \\ 0.604 & 0.531 & 0.130 \\ 0.656 & 0.594 & 0.260 \\ 0.614 & 0.870 & 0.187 \\ 0.376 & 0.234 & 0.377 \\ 0.500 & 0.226 & 0.476 \end{bmatrix}.$$

The standard series is taken to be the lowest level of the linguistic term describing all three risk factors, which is *Very Low*. When the linguistic term *Very Low* is defuzzified, the crisp number obtained is 0.130, this represents the average value, as such the value 0 (lowest possible value) is used to represent the linguistic term *Very Low* in the standard series (Liu et al. 2011). A matrix representing the standard series is generated as shown here

$$X_0 = \begin{bmatrix} H_{11} & H_{11} & H_{11} \\ H_{11} & H_{11} & H_{11} \\ H_{11} & H_{11} & H_{11} \\ H_{11} & H_{11} & H_{11} \\ H_{11} & H_{11} & H_{11} \\ H_{11} & H_{11} & H_{11} \\ H_{11} & H_{11} & H_{11} \end{bmatrix} = \begin{bmatrix} 0 & 0 & 0 \\ 0 & 0 & 0 \\ 0 & 0 & 0 \\ 0 & 0 & 0 \\ 0 & 0 & 0 \\ 0 & 0 & 0 \\ 0 & 0 & 0 \end{bmatrix}.$$

The difference between the comparative and the standard series D_0 is then calculated and expressed as a matrix. Since all entries for the matrix representing the standard series were determined to be 0, the difference between the comparative and the standard series would be equal to the comparative series (considering that $\Delta_{ij} = \left\|X_{0j} - X_{ij}\right\|$).

Using the values obtained from the difference of the standard and the comparative series, the grey relation coefficient, γ_{ij}, is calculated via Eq. (5.13) for each risk factor of the failure modes identified in the FMEA. Take the first failure mode in Table 5.5 for example, the grey relation coefficients for the risk factors O, S, and D are calculated as shown here:

$$\gamma_{11} = \frac{0.130 + 0.5 \times 0.870}{0.379 + 0.5 \times 0.870} = 0.694,$$
$$\gamma_{12} = \frac{0.130 + 0.5 \times 0.870}{0.364 + 0.5 \times 0.870} = 0.707,$$
$$\gamma_{13} = \frac{0.130 + 0.5 \times 0.870}{0.210 + 0.5 \times 0.870} = 0.876.$$

Similarly, the grey relation coefficients for all the failure modes with respect to each risk factor can be calculated in the same way as shown in the matrix below

$$\left[\gamma_{ij}\right]_{7\times 3} = \begin{bmatrix} 0.694 & 0.707 & 0.876 \\ 0.579 & 0.647 & 0.934 \\ 0.544 & 0.585 & 1.000 \\ 0.518 & 0.549 & 0.813 \\ 0.539 & 0.433 & 0.908 \\ 0.697 & 0.845 & 0.696 \\ 0.604 & 0.851 & 0.620 \end{bmatrix}.$$

On the other side, based upon the information in Table 5.3, the relative importance weights of risk factors are first aggregated using Eq. (5.1) as shown in the last row of Table 5.4. The group weights of risk factors are then defuzzified and normalized using Eqs. (5.7) and (5.2), respectively. The results are provided in the last row of Table 5.5.

Table 5.3 Assessment information on failure modes by the five FMEA team members (Liu et al. 2011)

Risk factors	Team members	Factor weights	Failure modes						
			FM1	FM2	FM3	FM4	FM5	FM6	FM7
O	TM1	M	H_{22}	(H_{33}, 0.50) (H_{44}, 0.50)	H_{44}	H_{44}	H_{44}	H_{22}	H_{33}
	TM2	H	H_{22}	H_{33}	(H_{33}, 0.90)	H_{44}	H_{44}	H_{22}	H_{33}
	TM3	M		H_{33}	H_{33}	H_{44}	H_{34}	H_{22}	H_{33}
	TM4	VH	H_{22}	H_{33}	H_{44}	H_{33}	H_{33}	(H_{22}, 0.20) (H_{35}, 0.80)	H_{33}
	TM5	M	H_{25}	H_{33}	H_{44}	H_{44}	H_{44}	H_{33}	H_{33}
S	TM1	VH	H_{22}	H_{33}	H_{44}	(H_{24}, 0.90)	H_{55}	H_{11}	(H_{11}, 0.70) (H_{22}, 0.30)
	TM2	H	H_{22}	H_{33}	H_{33}	H_{44}	H_{55}	H_{22}	H_{11}
	TM3	VH	(H_{12}, 0.50) (H_{34}, 0.50)	H_{22}	H_{33}	H_{33}	H_{55}	H_{11}	H_{22}
	TM4	H	H_{23}	H_{33}	H_{33}	H_{44}	H_{55}	(H_{13}, 0.80)	H_{22}
	TM5	H	H_{22}		H_{33}	H_{33}	H_{55}	H_{11}	H_{11}
D	TM1	L	H_{11}	H_{13}	H_{11}	H_{22}	H_{11}	H_{22}	H_{33}
	TM2	L	H_{12}	H_{11}	H_{11}	H_{11}	H_{11}	(H_{13}, 0.80) (H_{44}, 0.20)	H_{33}
	TM3	L	H_{11}	H_{11}	H_{11}	H_{22}	H_{11}	H_{22}	H_{23}
	TM4	M	H_{22}	H_{11}	H_{11}	H_{22}	H_{22}		(H_{12}, 0.40) (H_{33}, 0.60)
	TM5	L	(H_{12}, 0.95)	H_{11}	H_{11}	H_{22}	H_{22}	H_{22}	H_{33}

Table 5.4 Group assessments of the FMEA team on failure modes and group weights of risk factors (Liu et al. 2011)

Failure modes	*O*	*S*	*D*
FM1	{(H_{15}, 0.30), (H_{22}, 0.60), (H_{25}, 0.10)}	{(H_{12}, 0.15), (H_{22}, 0.45), (H_{23}, 0.25), (H_{34}, 0.15)}	{(H_{11}, 0.45), (H_{12}, 0.295), (H_{15}, 0.005), (H_{22}, 0.25)}
FM2	{(H_{33}, 0.975), (H_{44}, 0.075)}	{(H_{15}, 0.10), (H_{22}, 0.30), (H_{33}, 0.60)}	{(H_{11}, 0.85), (H_{13}, 0.15)}
FM3	{(H_{15}, 0.02), (H_{33}, 0.48), (H_{44}, 0.50)}	{(H_{33}, 0.85), (H_{44}, 0.15)}	{(H_{11}, 1.00)}
FM4	{(H_{33}, 0.25), (H_{44}, 0.75)}	{(H_{15}, 0.015), (H_{24}, 0.135), (H_{33}, 0.40), (H_{44}, 0.45)}	{(H_{11}, 0.20), (H_{22}, 0.80)}
FM5	{(H_{33}, 0.25), (H_{34}, 0.30), (H_{44}, 0.45)}	{(H_{55}, 1.00)}	{(H_{11}, 0.65), (H_{22}, 0.35)}
FM6	{(H_{22}, 0.70), (H_{33}, 0.10), (H_{35}, 0.20)}	{(H_{11}, 0.55), (H_{13}, 0.20), (H_{15}, 0.05), (H_{22}, 0.20)}	{(H_{13}, 0.16), (H_{15}, 0.25), (H_{22}, 0.55), (H_{44}, 0.04)}
FM7	{(H_{33}, 1.00)}	{(H_{11}, 0.405), (H_{22}, 0.595)}	{(H_{12}, 0.10), (H_{33}, 0.90)}
Group weights	(0.425, 0.675, 0.8625)	(0.6125, 0.8625, 1)	(0.0625, 0.2875, 0.5375)

Table 5.5 Defuzzified and aggregated assessment information for failure modes and risk priority ranking (Liu et al. 2011)

Failure modes	O	S	D	Γ_{ij}	Ranking
FM1	0.379	0.364	0.210	0.734	6
FM2	0.541	0.438	0.170	0.677	4
FM3	0.604	0.531	0.130	0.649	3
FM4	0.656	0.594	0.260	0.588	2
FM5	0.614	0.870	0.187	0.561	1
FM6	0.376	0.234	0.377	0.763	7
FM7	0.500	0.226	0.476	0.718	5
Weights	0.36	0.45	0.19		

Table 5.6 Defuzzified values for fuzzy assessment grades (Liu et al. 2011)

Assessment grades	Defuzzified values	Assessment grades	Defuzzified values	Assessment grades	Defuzzified values
H_{11}	0.130	H_{22}	0.292	H_{34}	0.567
H_{12}	0.259	H_{23}	0.433	H_{35}	0.606
H_{13}	0.394	H_{24}	0.500	H_{44}	0.708
H_{14}	0.459	H_{25}	0.541	H_{45}	0.741
H_{15}	0.500	H_{33}	0.500	H_{55}	0.870

Substituting the grey relation coefficients and group weights of risk factors into Eq. (5.14) will give the degree of relation for the first failure mode as seen here:

$$\begin{aligned}\Gamma_1 &= [(0.694 \times 0.36) + (0.707 \times 0.45) + (0.876 \times 0.19)] \\ &= 0.734.\end{aligned}$$

In the same way, the degrees of relation are calculated for all the failure modes identified in the FMEA to produce a ranking that determines the priority for attention. The results are shown in Table 5.5. The degrees of relation of the seven failure modes give the priority ranking of the seven failure modes as FM5 $\succ$ FM4 $\succ$ FM3 $\succ$ FM2 $\succ$FM7 $\succ$ FM1 $\succ$ FM6, which is perfectly consistent with the real-world situations of the failures in this study. So, the final conclusion for this example is that FM5 should be given the top priority for correction, followed by FM4, FM3, FM2, FM7, FM1, and FM6.

The potential applications of the proposed FMEA and the detailed computational process of the degree of relation are examined and illustrated with the above numerical example. The results show that the proposed FMEA provides a useful, practical, and flexible way for the risk evaluation in FMEA. In particular, the proposed FMEA model offered a new way for capturing MEA team members' opinions and prioritizing failure modes in FMEA. Compared with the conventional RPN method and its kinds of variants, the risk priority model here proposed has the following advantages: (1) The relative importance weights of risk factors are taken into consideration in the process of prioritization of failure modes, which makes the proposed FMEA more realistic, more practical and more flexible. (2) Risk factors and their relative importance weights are evaluated in a linguistic manner rather than in precise numerical values. This enables the domain experts to express their judgments more realistically and makes the assessment easier to be carried out. (3) The diversity and uncertainty of FMEA team members' assessment information can be well reflected and modeled using belief structures. And it provides an organized method to combine expert knowledge and experience for use in FMEA. (4) Failure modes can be fully ranked and well distinguished from each other unless some of them are assessed to be the same.

References

Chang CL, Liu PH, Wei CC (2001) Failure mode and effects analysis using grey theory. Integr Manuf Syst 12(3):211–216

Chen CB, Klein CM (1997) A simple approach to ranking a group of aggregated fuzzy utilities. IEEE Trans Syst Man Cybern B Cybern 27(1):26–35

Chen SM, Cheng SH, Chiou CH (2016) Fuzzy multiattribute group decision making based on intuitionistic fuzzy sets and evidential reasoning methodology. Inf Fusion 27:215–227

Deng JL (1989) Introduction to gray system theory. J Grey Syst 1(1):1–24

Fu C, Chin KS (2014) Robust evidential reasoning approach with unknown attribute weights. Knowl-Based Syst 59:9–20

Guo M, Yang JB, Chin KS, Wang HW, Liu XB (2009) Evidential reasoning approach for multiattribute decision analysis under both fuzzy and interval uncertainty. IEEE Trans Fuzzy Syst 17(3):683–697

Liu HC, Liu L, Bian QH, Lin QL, Dong N, Xu PC (2011) Failure mode and effects analysis using fuzzy evidential reasoning approach and grey theory. Expert Syst Appl 38(4):4403–4415

Liu HC, Liu L, Liu N (2013) Risk evaluation approaches in failure mode and effects analysis: a literature review. Expert Syst Appl 40(2):828–838

Liu HC, Li P, You JX, Chen YZ (2015) A novel approach for FMEA: combination of interval 2-tuple linguistic variables and grey relational analysis. Qual Reliab Eng Int 31(5):761–772

Shafer G (1976) A mathematical theory of evidence. Princeton University Press, New Jersey

Wang YM, Chin KS, Poon GKK, Yang JB (2009) Risk evaluation in failure mode and effects analysis using fuzzy weighted geometric mean. Expert Syst Appl 36(2):1195–1207

Yang JB, Singh MG (1994) An evidential reasoning approach for multiple-attribute decision making with uncertainty. IEEE Trans Syst Man Cybern 24(1):1–18

Yang JB, Wang YM, Xu DL, Chin KS (2006) The evidential reasoning approach for MADA under both probabilistic and fuzzy uncertainties. Eur J Oper Res 171(1):309–343

Chapter 6
FMEA Using D Numbers and Grey Relational Projection Method

As stated in Chap. 5, two critical issues of FMEA are the representation and handling of various types of risk assessments and the determination of risk priorities of failure modes. In Liu et al. (2014), a new representation of uncertain information, called D numbers, was introduced to handle different assessments of risk factors provided by FMEA team members. An improved grey relational analysis method, grey relational projection (GRP), was used to determine the risk priority order of the failure modes that have been identified. Based on D numbers and GRP method, a new risk priority model was then proposed for the risk evaluation in FMEA. The new model can not only effectively deal with the various uncertainties in the risk assessment process but also rank the risk of the identified failure modes in a comprehensive way. Moreover, the proposed model overcomes the deficiencies surrounding the conventional RPN method and provides a new framework for prioritizing failure modes in FMEA.

6.1 D Numbers

D number (Deng 2012; Deng et al. 2014b) is a new representation of uncertain information, which is introduced to overcome the shortcomings existed in Dempster–Shafer (D–S) theory (Dempster 1967; Shafer 1976).

Definition 6.1 (Deng 2012; Deng et al. 2014b) Let Ω be a finite nonempty set, D number is a mapping formulated by

$$D : \Omega \to [0, 1] \quad (6.1)$$

H.-C. Liu, *FMEA Using Uncertainty Theories and MCDM Methods*,
DOI 10.1007/978-981-10-1466-6_6

with

$$\sum_{B \subseteq \Omega} D(B) \leq 1 \quad \text{and} \quad D(\varnothing) = 0 \tag{6.2}$$

where $\varnothing$ is an empty set and B is a subset of Ω. It is worth pointing out that different from the concept of frame of discernment in D–S theory, the elements in set Ω do not require mutually exclusive and the completeness constraint is released in D numbers. If $\sum_{B \subseteq \Omega} D(B) = 1$, the information is said to be complete; otherwise, the information is assumed to be incomplete.

For a discrete set $\Omega = \{b_1, b_2, \ldots, b_i, \ldots, b_n\}$, where $b_i \in R$ and $b_i \neq b_j$ if $i \neq j$, a special form of D numbers can be expressed by

$$\begin{aligned} D(\{b_1\}) &= v_1 \\ D(\{b_2\}) &= v_2 \\ &\cdots \\ D(\{b_i\}) &= v_i \\ &\cdots \\ D(\{b_n\}) &= v_n \end{aligned} \tag{6.3}$$

or simply denoted as $D = \{(b_1, v_1), (b_2, v_2), \ldots, (b_i, v_i), \ldots, (b_n, v_n)\}$, where $v_i > 0$ and $\sum_{i=1}^{n} v_i \leq 1$.

Some properties of D numbers are introduced as follows.

Definition 6.2 (*Permutation invariability*) (Deng et al. 2014a, b) If there are two D numbers that

$$D_1 = \{(b_1, v_1), \ldots, (b_i, v_i), \ldots, (b_n, v_n)\}$$

and

$$D_2 = \{(b_n, v_n), \ldots, (b_i, v_i), \ldots, (b_1, v_1)\},$$

then $D_1 \Leftrightarrow D_2$.

Definition 6.3 (Deng 2012; Deng et al. 2014b) For $D = \{(b_1, v_1), (b_2, v_2), \ldots, (b_i, v_i), \ldots, (b_n, v_n)\}$, the integration representation of D is defined as

$$I(D) = \sum_{i=1}^{n} b_i v_i \tag{6.4}$$

where $v_i > 0$ and $\sum_{i=1}^{n} v_i \leq 1$.

Next, the combination rule for D numbers is given $b_i \in R$ as below:

Definition 6.4 (Deng 2012; Deng et al. 2014a) Let D_1 and D_2 be two D numbers that:

$$D_1 = \{(b_1^1, v_1^1), \ldots, (b_i^1, v_i^1), \ldots, (b_n^1, v_n^1)\}$$

$$D_2 = \left\{(b_1^2, v_1^2), \ldots, \left(b_j^2, v_j^2\right), \ldots, (b_m^2, v_m^2)\right\}$$

Then, the combination of D_1 and D_2, expressed as $D = D_1 \oplus D_2$, is defined by

$$D(b) = v \tag{6.5}$$

with

$$b = \frac{b_i^1 + b_j^2}{2} \tag{6.6}$$

$$v = \frac{v_i^1 + v_j^2}{2} \Big/ C \tag{6.7}$$

$$C = \begin{cases} \sum_{j=1}^{m}\sum_{i=1}^{n}\left(\frac{v_i^1 + v_j^2}{2}\right), & \sum_{i=1}^{n} v_i^1 = 1 \text{ and } \sum_{j=1}^{m} v_j^2 = 1; \\ \sum_{j=1}^{m}\sum_{i=1}^{n}\left(\frac{v_i^1 + v_j^2}{2}\right) + \sum_{j=1}^{m}\left(\frac{v_c^1 + v_j^2}{2}\right), & \sum_{i=1}^{n} v_i^1 < 1 \text{ and } \sum_{j=1}^{m} v_j^2 = 1; \\ \sum_{j=1}^{m}\sum_{i=1}^{n}\left(\frac{v_i^1 + v_j^2}{2}\right) + \sum_{j=1}^{m}\left(\frac{v_i^1 + v_c^2}{2}\right), & \sum_{i=1}^{n} v_i^1 = 1 \text{ and } \sum_{j=1}^{m} v_j^2 < 1; \\ \sum_{j=1}^{m}\sum_{i=1}^{n}\left(\frac{v_i^1 + v_j^2}{2}\right) + \sum_{j=1}^{m}\left(\frac{v_c^1 + v_j^2}{2}\right) & \\ \quad + \sum_{j=1}^{m}\left(\frac{v_i^1 + v_c^2}{2}\right) + \frac{v_c^1 + v_c^2}{2}, & \sum_{i=1}^{n} v_i^1 < 1 \text{ and } \sum_{j=1}^{m} v_j^2 < 1. \end{cases} \tag{6.8}$$

where $v_c^1 = 1 - \sum_{i=1}^{n} v_i^1$ and $v_c^2 = 1 - \sum_{j=1}^{m} v_j^1$.

Note that the combination operation defined in Definition 6.4 does not preserve the associative property (Deng et al. 2014a). That is $(D_1 \oplus D_2) \oplus D_3 \neq D_1 \oplus (D_2 \oplus D_3) \neq (D_1 \oplus D_3) \oplus D_2$. Therefore, a combination operation for multiple D numbers has been developed in order that multiple D numbers can be combined correctly and efficiently.

Definition 6.5 (Deng et al. 2014a) Let D_1, D_2, …, D_n be n D numbers, μ_j is an order variable for each D_j, indicated by tuple $\left\langle \mu_j, D_{\mu_j} \right\rangle$, then the combination operation of multiple D numbers is a mapping f_D, such that

$$f_D = (D_1, D_2, \ldots, D_n) = [\cdots [D_{\lambda_1} \oplus D_{\lambda_2}] \oplus \cdots \oplus D_{\lambda_n}] \quad (6.9)$$

where D_{λ_i} is the D_{μ_i} of the tuple $\left\langle \mu_j, D_{\mu_j} \right\rangle$ having the ith lowest μ_j.

6.2 The Proposed Model for FMEA

The grey theory (Deng 1989) is an effective mathematical tool to deal with the systems characterized by poor, incomplete, and uncertain information. Based on grey relational analysis (GRA) method and vector projection, grey relational projection (GRP) method is developed (Zheng et al. 2010; Zhang et al. 2013). The major advantages of the GRP method are that the results are based on original data and the calculation is simple and reliable. The flowchart in Fig. 6.1 shows the proposed model based on *D* numbers and GRP method to prioritize the individuated failure modes in FMEA. The *D* numbers are used to deal with and model the diversity and uncertainty of evaluation information provided by FMEA team members. After aggregating individual evaluation information into group assessments, the GRP method is applied to the prioritization and selection of failure modes. Particularly, both positive and negative reference sequences are considered in the determination of risk priorities of failure modes.

Supposing there are l cross-functional members, TM_k $(k = 1, \ldots, l)$, in a FMEA team responsible for the assessment of m failure modes, FM_i $(i = 1, \ldots, m)$, with respect to n risk factors, RF_j $(j = 1, \ldots, n)$. Each team member TM_k is given a weight $\lambda_k > 0$ $(k = 1, \ldots, l)$ satisfying $\Sigma_{k=1}^{l} \lambda_k = 1$ to reflect his/her relative importance in the failure analysis process. The steps of the proposed FMEA model are given as follows (Liu et al. 2014):

Step 1. Identify the objectives of risk assessment and define the risk analysis level

The first step is defining the objectives of risk assessment and defining the risk analysis level. Giving clear and careful thought to this step is very critical to the following risk determining process.

Step 2. Arrange a FMEA team, list all potential failure modes and describe a finite set of relevant risk factors

Application of FMEA is a group decision function and cannot be accomplished on an individual basis. Thus, a cross-functional and multidisciplinary team should be established for listing all the potential failure modes of a specific product or system and described a set of relevant risk factors.

Step 3. Evaluate the failure modes and risk factor weights using D numbers

Considering their personal backgrounds and different expertise, the FMEA team members may express different judgments for the risk factors and their importance weights, which inevitably involves uncertainty and incompleteness. For this reason, the assessments for risk factors and their relative weights can be implemented by using D numbers. But before making the evaluations, appropriate numeric scales

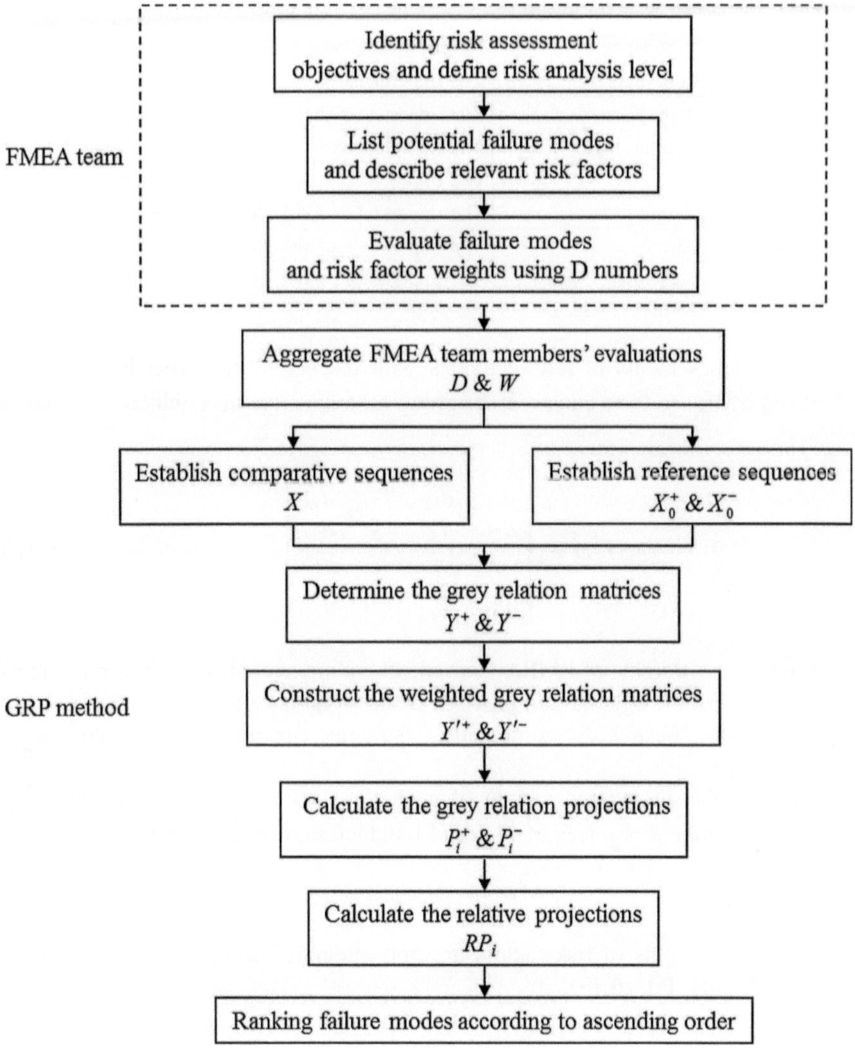

Fig. 6.1 Flowchart of the proposed FMEA model (Liu et al. 2014)

(ratings) should be defined first in the failure analysis. For example, the 10-point scales shown in Tables 1.1–1.3 can be adopted for evaluating the failure modes with respect to each risk factor and the 7-point scale shown in Table 6.1 can be used for assessing the relative importance of risk factors.

Step 4. Aggregate the FMEA team members' individual evaluations

Let d_{ij}^k be the D number provided by TM_k on the assessment of FM_i with respect to RF_j, and w_j^k be the weight of risk factor RF_j given by TM_k to reflect its relative importance in the determination of risk priorities of the failure modes. Then, the

Table 6.1 Suggested scale for importance of risk factors (Liu et al. 2014)

Rating	Importance	Description
7	Very high	The importance of risk factor is very high
6	High	The importance of risk factor is high
5	Medium high	The importance of risk factor is medium high
4	Medium	The importance of risk factor is medium
3	Medium low	The importance of risk factor is medium low
2	Low	The importance of risk factor is low
1	Very low	The importance of risk factor is very low

aggregated assessments of failure modes with respect to each risk factor d_{ij} are calculated by Eq. (6.9) to construct the group assessment matrix, which is shown as follows:

$$D = \begin{bmatrix} d_{11} & d_{12} & \cdots & d_{1n} \\ d_{21} & d_{22} & \cdots & d_{2n} \\ \vdots & \vdots & \vdots & \vdots \\ d_{m1} & d_{m2} & \cdots & d_{mn} \end{bmatrix}. \tag{6.10}$$

Note that the weights of FMEA team members are treated as the order variables (i.e., $\mu_k = \lambda_k$) in the aggregation process. If the weights of some team member are equal, then the "best–worst combination" strategy can be adopted (Deng et al. 2014a).

Similarly, the aggregated weight of each risk factor w_j can be calculated by Eq. (6.9) to get the group weight vector of risk factors, which is denoted as follows:

$$W = (w_1, w_2, \ldots, w_n). \tag{6.11}$$

The group weights of risk factors are first integrated using Eq. (6.4) and then normalized using Eq. (6.12).

$$\bar{w}_j = \frac{I(w_j)}{\sum_{j=1}^{n} I(w_j)}, \tag{6.12}$$

where $I(w_j)$ is referred to as the integration representation of the group risk factor weight determined by Eq. (6.4).

Step 5. Establish the comparative sequences

An information sequence with n components or decision factors can be expressed as follows: $X_i' = (x_{i1}', x_{i2}', \ldots, x_{in}') \in X$, where x_{ij}' denotes the jth factor of X_i'. If all information sequences are comparable, the n information sequence can be described as the following matrix:

$$X = \begin{bmatrix} x_{11} & x_{12} & \cdots & x_{1n} \\ x_{21} & x_{22} & \cdots & x_{2n} \\ \vdots & \vdots & \vdots & \vdots \\ x_{m1} & x_{m2} & \cdots & x_{mn} \end{bmatrix}. \tag{6.13}$$

For the application of this matrix to FMEA, the values of x_{ij} represent the integration representation of the aggregated assessment d_{ij} in the group assessment matrix D by using Eq. (6.4), i.e., $x_{ij} = I(d_{ij})$.

Step 6. Establish the reference sequences

In this study, the GRP established on double base points (the positive-ideal alternative and the negative-ideal alternative) is employed; thus, the positive reference sequence and the negative reference sequence should be determined and expressed as $X_0^+ = (x_{01}^+, x_{02}^+, \ldots, x_{0n}^+)$ and $X_0^- = (x_{01}^-, x_{02}^-, \ldots, x_{0n}^-)$, respectively. When conducting FMEA, the lowest levels of all the risk factors are desired; therefore, the two reference sequences for risk factors can be defined as follows:

$$X_0^+ = (1, 1, \ldots, 1), \tag{6.14}$$

$$X_0^- = (10, 10, \ldots, 10). \tag{6.15}$$

Step 7. Determine the grey relation matrices

The grey relation coefficient between x_{ij} and x_{0j}^+ is calculated by the following formula

$$\gamma_{ij}^+ = \frac{\min\limits_{1 \le i \le m} \min\limits_{1 \le j \le n} \left| x_{0j}^+ - x_{ij} \right| + \zeta \max\limits_{1 \le i \le m} \max\limits_{1 \le j \le n} \left| x_{0j}^+ - x_{ij} \right|}{\left| x_{0j}^+ - x_{ij} \right| + \zeta \max\limits_{1 \le i \le m} \max\limits_{1 \le j \le n} \left| x_{0j}^+ - x_{ij} \right|}, \tag{6.16}$$

where γ_{ij}^+ is the grey relation coefficient of x_{ij} with respect to the positive-ideal risk factor x_{0j}^+, ζ is the distinguishing coefficient, and $\zeta \in [0, 1]$. Generally, $\zeta = 0.5$ is applied.

Similarly, the grey relation coefficient between x_{ij} and x_{0j}^- is calculated by Eq. (6.17).

$$\gamma_{ij}^- = \frac{\min\limits_{1 \le i \le m} \min\limits_{1 \le j \le n} \left| x_{0j}^- - x_{ij} \right| + \zeta \max\limits_{1 \le i \le m} \max\limits_{1 \le j \le n} \left| x_{0j}^- - x_{ij} \right|}{\left| x_{0j}^- - x_{ij} \right| + \zeta \max\limits_{1 \le i \le m} \max\limits_{1 \le j \le n} \left| x_{0j}^- - x_{ij} \right|}, \tag{6.17}$$

where γ_{ij}^- is the grey relation coefficient of x_{ij} with respect to the negative-ideal risk factor x_{0j}^-.

Based on γ_{ij}^{+} and γ_{ij}^{-}, the grey relation matrices can be determined as follows:

$$Y^{+} = \begin{bmatrix} \gamma_{11}^{+} & \gamma_{12}^{+} & \cdots & \gamma_{1n}^{+} \\ \gamma_{21}^{+} & \gamma_{22}^{+} & \cdots & \gamma_{2n}^{+} \\ \vdots & \vdots & \vdots & \vdots \\ \gamma_{m1}^{+} & \gamma_{m2}^{+} & \cdots & \gamma_{mn}^{+} \end{bmatrix}, \tag{6.18}$$

$$Y^{-} = \begin{bmatrix} \gamma_{11}^{-} & \gamma_{12}^{-} & \cdots & \gamma_{1n}^{-} \\ \gamma_{21}^{-} & \gamma_{22}^{-} & \cdots & \gamma_{2n}^{-} \\ \vdots & \vdots & \vdots & \vdots \\ \gamma_{m1}^{-} & \gamma_{m2}^{-} & \cdots & \gamma_{mn}^{-} \end{bmatrix}, \tag{6.19}$$

where Y^{+} is the grey relation matrix between every failure mode and the positive reference sequence; Y^{-} is the grey relation matrix between every failure mode and the negative reference sequence.

Step 8. Construct the weighted grey relation matrices

The weighted grey relation matrices Y'^{+} and Y'^{-} are obtained using Eqs. (6.20) and (6.21) as follows:

$$Y'^{+} = \begin{bmatrix} \bar{w}_1\gamma_{11}^{+} & \bar{w}_2\gamma_{12}^{+} & \cdots & \bar{w}_n\gamma_{1n}^{+} \\ \bar{w}_1\gamma_{21}^{+} & \bar{w}_2\gamma_{22}^{+} & \cdots & \bar{w}_n\gamma_{2n}^{+} \\ \vdots & \vdots & \vdots & \vdots \\ \bar{w}_1\gamma_{m1}^{+} & \bar{w}_2\gamma_{m2}^{+} & \cdots & \bar{w}_n\gamma_{mn}^{+} \end{bmatrix}, \tag{6.20}$$

$$Y'^{-} = \begin{bmatrix} \bar{w}_1\gamma_{11}^{-} & \bar{w}_2\gamma_{12}^{-} & \cdots & \bar{w}_n\gamma_{1n}^{-} \\ \bar{w}_1\gamma_{21}^{-} & \bar{w}_2\gamma_{22}^{-} & \cdots & \bar{w}_n\gamma_{2n}^{-} \\ \vdots & \vdots & \vdots & \vdots \\ \bar{w}_1\gamma_{m1}^{-} & \bar{w}_2\gamma_{m2}^{-} & \cdots & \bar{w}_n\gamma_{mn}^{-} \end{bmatrix}, \tag{6.21}$$

where $\bar{w}_j$ is the normalized group weight of the risk factor RF_j.

Step 9. Calculate the grey relational projections

Each line in the weighted grey relation matrix Y'^{+} can be considered as a row vector, which denotes the corresponding failure mode. Therefore, the grey relational projection of the failure mode FM_i on the positive reference sequence X_0^{+} is calculated as follows:

$$P_i^{+} = \|Y_i'^{+}\| \cos\left(Y_i'^{+}, Y_0'^{+}\right) = \sqrt{\sum_{j=1}^{n}\left(\bar{w}_j\gamma_{ij}^{+}\right)^2} \times \frac{\sum_{j=1}^{n}\left(\left(\bar{w}_j\gamma_{ij}^{+}\right) \times \bar{w}_j\right)}{\sqrt{\sum_{j=1}^{n}\left(\bar{w}_j\gamma_{ij}^{+}\right)^2} \times \sqrt{\sum_{j=1}^{n}\bar{w}_j^2}}$$
$$= \sum_{j=1}^{n}\left(\frac{\bar{w}_j^2}{\sqrt{\sum_{j=1}^{n}\bar{w}_j^2}} \times \gamma_{ij}^{+}\right). \tag{6.22}$$

where $Y_0'^{+} = \left(\bar{w}_1\gamma_{01}^{+}, \bar{w}_2\gamma_{02}^{+}, \ldots, \bar{w}_n\gamma_{0n}^{+}\right) = (\bar{w}_1, \bar{w}_2, \ldots, \bar{w}_n)$ is the weighted grey relation coefficient between positive reference sequences.

Similarly, the grey relational projection of the failure mode FM_i on the negative reference sequence X_0^- is calculated as follows:

$$P_i^- = \left\|Y_i'^-\right\| \cos\left(Y_i'^-, Y_0'^-\right) = \sqrt{\sum_{j=1}^{n}\left(\bar{w}_j\gamma_{ij}^-\right)^2} \times \frac{\sum_{j=1}^{n}\left(\left(\bar{w}_j\gamma_{ij}^-\right) \times \bar{w}_j\right)}{\sqrt{\sum_{j=1}^{n}\left(\bar{w}_j\gamma_{ij}^-\right)^2} \times \sqrt{\sum_{j=1}^{n}\bar{w}_j^2}}$$

$$= \sum_{j=1}^{n}\left(\frac{\bar{w}_j^2}{\sqrt{\sum_{j=1}^{n}\bar{w}_j^2}} \times \gamma_{ij}^-\right). \tag{6.23}$$

where $Y_0'^{-} = \left(\bar{w}_1\gamma_{01}^{-}, \bar{w}_2\gamma_{02}^{-}, \ldots, \bar{w}_n\gamma_{0n}^{-}\right) = (\bar{w}_1, \bar{w}_2, \ldots, \bar{w}_n)$ is the weighted grey relation coefficient between negative reference sequences.

Let the weight of grey relational projection be $w_j' = \frac{\bar{w}_j^2}{\sqrt{\sum_{j=1}^{n}\bar{w}_j^2}}$, then

$$P_i^+ = \sum_{j=1}^{n}\left(w_j' \times \gamma_{ij}^+\right), \tag{6.24}$$

$$P_i^- = \sum_{j=1}^{n}\left(w_j' \times \gamma_{ij}^-\right). \tag{6.25}$$

Step 10. Calculate the relative projection and rank the failure modes

The relative projection (RP) of each failure mode to positive reference sequence is defined as follows:

$$\text{RP}_i = \frac{P_i^+}{P_i^+ + P_i^-}. \tag{6.26}$$

In FMEA, the relative projection value denotes the relationship between the potential failure mode and the optimal value of risk factors. The higher the value of relative projection obtained from Eq. (6.26), the smaller the effect of the identified failure mode. Thus, all failure modes in FMEA can be prioritized according to the ascending order of their relative projection coefficients.

6.3 An Illustrative Example

6.3.1 Implementation

In this section, the proposed FMEA model has been applied to a case of rotor blades for an aircraft turbine (Yang et al. 2011; Liu et al. 2014), to illustrate the potentials of using D numbers and GRP method in capturing FMEA team members' diversity evaluations and ranking failure modes under different types of uncertainties. Rotor blades are the key rotating component of an aircraft turbine. Since they are the thin-form components moving in high-speed rotation, under the severe load conditions in complex work environments, rotor blades are the components that are most likely to be failed in aircraft turbines. Meanwhile, with the development of the aviation industry, the thrust–weight ratio (TWR) of aircraft turbines has become higher and higher. The stress level of rotor blades is greatly increasing as well. Thus, their reliability plays an important role in the aircraft turbine security. In order to improve the safety and reliability of rotor blades, FMEA is prerequisite in their design.

The rotor blades consist of two different subsystems: the compressor rotor blades and the turbo rotor blades. The FMEA in this example is limited only to the subsystem of the compressor rotor blades and the failure modes which could lead to an accident with undesired consequences are considered. For each of the failure modes, the system is investigated for any alarm or condition monitoring arrangement in accordance with the practical engineering background. As a result, there are eight major potential failure modes ($FM_i, i = 1, 2, \ldots, 8$) which were identified by a FMEA team and need to be prioritized so that high risky failure modes can be corrected with top priorities. Suppose the FMEA team is made up of three experts, TM_1, TM_2, and TM_3, each evaluates the failure modes in terms of the risk factors O, S, and D. The FMEA team members evaluate the risk factors and their relative importance weights using D numbers based on the numeric ratings defined in Tables 1.1–1.3 and Table 6.1. The assessment results of the three experts on the eight failure modes and the risk factor weights are presented in Table 6.2. To reflect their differences in performing FMEA, the three team members are assigned the following weights, 0.25, 0.40, and 0.35, because of their different domain knowledge and expertise.

For carrying out the risk analysis using the proposed model, we first synthesize individual assessments of the FMEA team members into group assessments by using Eq. (6.9), as shown in Table 6.3. The comparative sequences are then generated using Eq. (6.4), and the results are tabulated in Table 6.4.

The positive and negative reference sequences are taken to be the lowest and highest levels of the risk factors, respectively. The matrices representing the two reference sequences are generated as shown here:

Table 6.2 Evaluation information on failure modes provided by the FMEA team members (Liu et al. 2014)

Team members		TM1			TM2			TM3		
Risk factors		O	S	D	O	S	D	O	S	D
Failure modes	FM1	{(3, 0.4), (4, 0.6)}	{(7, 0.8)}	{(2, 1.0)}	{(3, 0.9), (4, 0.1)}	{(7, 1.0)}	{(2, 1.0)}	{(3, 0.8), (4, 0.2)}	{(7, 1.0)}	{(2, 1.0)}
	FM2	{(2, 1.0)}	{(8, 1.0)}	{(4, 1.0)}	{(2, 1.0)}	{(8, 0.7), (9, 0.3)}	{(4, 1.0)}	{(2, 1.0)}	{(8, 1.0)}	{(4, 1.0)}
	FM3	{(1, 0.9)}	{(10, 1.0)}	{(3, 1.0)}	{(1, 1.0)}	{(10, 1.0)}	{(3, 1.0)}	{(1, 1.0)}	{(10, 1.0)}	{(3, 0.9)}
	FM4	{(1, 1.0)}	{(6, 0.8), (7, 0.1)}	{(3, 1.0)}	{(1, 1.0)}	{(6, 1.0)}	{(2, 0.3), (3, 0.7)}	{(1, 1.0)}	{(6, 1.0)}	{(3, 1.0)}
	FM5	{(1, 1.0)}	{(3, 1.0)}	{(1, 0.5), (2, 0.5)}	{(1, 1.0)}	{(3, 1.0)}	{(1, 0.7), (2, 0.3)}	{(1, 1.0)}	{(2, 0.3), (3, 0.6)}	{(1, 1.0)}
	FM6	{(2, 1.0)}	{(6, 1.0)}	{(5, 1.0)}	{(2, 1.0)}	{(6, 1.0)}	{(5, 1.0)}	{(2, 1.0)}	{(6, 1.0)}	{(5, 1.0)}
	FM7	{(1, 0.7)}	{(7, 1.0)}	{(3, 1.0)}	{(1, 1.0)}	{(7, 0.8), (8, 0.1)}	{(3, 1.0)}	{(1, 1.0)}	{(7, 1.0)}	{(3, 1.0)}
	FM8	{(3, 1.0)}	{(5, 0.6), (6, 0.4)}	{(1, 1.0)}	{(3, 1.0)}	{(5, 0.75), (6, 0.2)}	{(1, 1.0)}	{(3, 1.0)}	{(5, 0.8), (7, 0.2)}	{(1, 1.0)}
Weights		{(6, 1.0)}	{(7, 1.0)}	{(5, 1.0)}	{(7, 1.0)}	{(7, 1.0)}	{(5, 1.0)}	{(6, 1.0)}	{(7, 1.0)}	{(5, 1.0)}

Table 6.3 Group assessments of FMEA team members and group weights of risk factors (Liu et al. 2014)

Failure modes	O	S	D
FM1	{(3, 0.3), (3.5, 0.5), (4, 0.2)}	{(7, 0.533)}	{(2, 1.0)}
FM2	{(2, 1.0)}	{(8, 0.567), (8.5, 0.433)}	{(4, 1.0)}
FM3	{(1, 0.544)}	{(10, 1.0)}	{(3, 0.544)}
FM4	{(1, 1.0)}	{(6, 0.4), (6.25, 0.335)}	{(2.5, 0.433), (3, 0.567)}
FM5	{(1, 1.0)}	{(2.75, 0.331), (3, 0.35)}	{(1, 0.3), (1.25, 0.3), (1.5, 0.2), (1.75, 0.2)}
FM6	{(2, 1.0)}	{6, 1.0)}	{(5, 1.0)}
FM7	{(1, 0.522)}	{(7, 0.6), (7.5, 0.367)}	{(3, 1.0)}
FM8	{(3, 1.0)}	{(5, 0.183), (5.25, 0.175), (5.5, 0.25), (5.75, 0.233), (6, 0.067), (6.25, 0.058)}	{(1, 1.0)}
Weights	{(6.75, 1.0)}	{(7, 1.0)}	{(5, 1.0)}

Table 6.4 Comparative sequences for the failure modes (Liu et al. 2014)

Failure modes	O	S	D
FM1	6.95	3.731	2
FM2	2	8.217	4
FM3	0.544	10	1.632
FM4	1	4.494	2.784
FM5	1	1.963	1.325
FM6	2	6	5
FM7	0.522	6.955	3
FM8	3	5.313	1
Weights	0.360	0.373	0.267

$$X_0^+ = \begin{bmatrix} 1 & 1 & 1 \\ 1 & 1 & 1 \\ 1 & 1 & 1 \\ 1 & 1 & 1 \\ 1 & 1 & 1 \\ 1 & 1 & 1 \\ 1 & 1 & 1 \\ 1 & 1 & 1 \end{bmatrix}, \quad X_0^- = \begin{bmatrix} 10 & 10 & 10 \\ 10 & 10 & 10 \\ 10 & 10 & 10 \\ 10 & 10 & 10 \\ 10 & 10 & 10 \\ 10 & 10 & 10 \\ 10 & 10 & 10 \\ 10 & 10 & 10 \end{bmatrix}.$$

According to Eqs. (6.16)–(6.19), the grey relation matrices Y^+ and Y^- are calculated, which are expressed as below:

Table 6.5 Results of the GRP method and risk priority ranking (Liu et al. 2014)

Failure modes	P_i^+	P_i^-	RP_i	Ranking
FM1	0.344	0.283	0.548	3
FM2	0.347	0.310	0.528	2
FM3	0.389	0.357	0.521	1
FM4	0.444	0.235	0.653	7
FM5	0.533	0.208	0.719	8
FM6	0.360	0.272	0.570	4
FM7	0.388	0.269	0.591	5
FM8	0.398	0.252	0.612	6

$$Y^+ = \begin{bmatrix} 0.431 & 0.622 & 0.818 \\ 0.818 & 0.384 & 0.600 \\ 0.908 & 0.333 & 0.877 \\ 1.000 & 0.563 & 0.716 \\ 1.000 & 0.824 & 0.933 \\ 0.818 & 0.474 & 0.529 \\ 0.904 & 0.430 & 0.692 \\ 0.692 & 0.511 & 1.000 \end{bmatrix}, \quad Y^- = \begin{bmatrix} 0.608 & 0.431 & 0.372 \\ 0.372 & 0.727 & 0.441 \\ 0.334 & 1.000 & 0.362 \\ 0.345 & 0.463 & 0.396 \\ 0.345 & 0.371 & 0.353 \\ 0.372 & 0.542 & 0.487 \\ 0.333 & 0.609 & 0.404 \\ 0.404 & 0.503 & 0.345 \end{bmatrix}.$$

On the other side, based upon the information in Table 6.2, the relative weights of risk factors are first aggregated using Eq. (6.9) as shown in the last row of Table 6.3. The group weights of risk factors are then integrated and normalized using Eqs. (6.4) and (6.12) successively. The calculation results are provided in the last row of Table 6.4.

Using the grey relation coefficients and the normalized group weights of risk factors, the grey relational projections, P_i^+ and P_i^-, are calculated by Eqs. (6.22) and (6.23) for all the failure modes identified in the FMEA. Finally, the relative projection of each failure mode to positive reference sequence, RP_i, can be calculated using Eq. (6.26). The results so obtained and the risk priority ranking of the eight failure modes are shown in Table 6.5. As is clear from Table 6.5, FM_3 has the smallest relative projection value in the failure modes of compressor rotor blades and thus should be given a top risk priority, followed by FM_2, FM_1, FM_6, FM_7, FM_8, FM_4, and FM_5. Therefore, the priority ranking of the eight failure modes is $\mathrm{FM}_3 \succ \mathrm{FM}_2 \succ \mathrm{FM}_1 \succ \mathrm{FM}_6 \succ \mathrm{FM}_7 \succ \mathrm{FM}_8 \succ \mathrm{FM}_4 \succ \mathrm{FM}_5$.

6.3.2 Comparisons and Discussion

In the previous literature (Yang et al. 2011), a risk evaluation method using D–S theory was proposed and the risk priority ranking obtained by this method is $\mathrm{FM}_2 \succ \mathrm{FM}_6 \succ \mathrm{FM}_1 \succ \mathrm{FM}_3 \succ \mathrm{FM}_7 \succ \mathrm{FM}_4 \succ \mathrm{FM}_8 \succ \mathrm{FM}_5$. Comparing the results obtained for the FMEA using the proposed approach and the method of Yang et al.

(2011), it can be found that except for FM_1, FM_5, and FM_7 the ranking orders of the other five failure modes are different. This may be explained by the fact that the relative importance among O, S, and D is not taken into account in the method suggested by Yang et al. (2011). Moreover, when multiple experts give same and precise values of the risk factors, the risk ranking obtained by the Yang et al.'s (2011) approach corresponds with the one obtained by the conventional RPN method. That is, this approach could not solve the shortcomings of the traditional FMEA. For example, FM_3 turned out to the most critical failure in terms of the proposed method, which has a higher priority compared to FM_2. After conducting criticality assessment using the method of Yang et al. (2011), FM_3 ranks only at the fourth place. At the same time, FM_2 becomes the most critical one. However, a close look at the values of the risk factors for FM_2 and FM_3 reveals that FM_3 has the highest value of S, which is the most important risk factor in comparison with O and D. Thus, giving FM_3 as the first priority which is obtained by the proposed method seems more genuine than that given by Yang et al. (2011). Following the similar logic and keeping in view of the weights of the three risk factors, ranking FM_8 as the sixth place by the proposed FMEA is more reasonable than ranking FM_4 as the sixth place by Yang et al. (2011). Therefore, the proposed method is more logical and a more accurate ranking can be achieved. Furthermore, the proposed model based on D numbers can effectively deal with various uncertainties, such as imprecision, fuzziness, and incompleteness, in the failure analysis process, which overcomes the shortcomings of the D–S theory.

The comparison analysis shows that a more accurate, reasonable risk ranking can be achieved by the combination of D numbers and GRP method for the risk evaluation in FMEA. In summary, the main advantages of the proposed risk priority model are as follows. First, the relative importance weights of risk factors are taken into consideration in the process of prioritization of failure modes. Second, various uncertainties in the assessments of FMEA team members, i.e., fuzziness, incompleteness, and imprecision, can be well reflected and modeled using D numbers. Third, the proposed model can solve the problem of discrete ordinal measurement scale and simple multiplication operation, which may cause meaningless and misleading results. Fourth, the proposed FMEA can get a more accurate risk ranking than the conventional RPN and other methods by using the double-reference-point-based GRP method.

References

Dempster A (1967) Upper and lower probabilities induced by a multivalued mapping. Ann Math Stat 38:325–329

Deng JL (1989) Introduction to gray system theory. J Grey Syst 1(1):1–24

Deng X, Hu Y, Deng Y, Mahadevan S (2014a) Environmental impact assessment based on D numbers. Expert Syst Appl 41(2):635–643

Deng X, Hu Y, Deng Y, Mahadevan S (2014b) Supplier selection using AHP methodology extended by D numbers. Expert Syst Appl 41(1):156–167

Deng Y (2012) D numbers: theory and applications. J Inf Comput Sci 9(9):2421–2428
Liu HC, You JX, Fan XJ, Lin QL (2014) Failure mode and effects analysis using D numbers and grey relational projection method. Expert Syst Appl 41(10):4670–4679
Shafer G (1976) A mathematical theory of evidence. Princeton University Press, New Jersey
Yang JP, Huang HZ, He LP, Zhu SP, Wen DW (2011) Risk evaluation in failure mode and effects analysis of aircraft turbine rotor blades using Dempster-Shafer evidence theory under uncertainty. Eng Fail Anal 18(8):2084–2092
Zhang X, Jin F, Liu P (2013) A grey relational projection method for multi-attribute decision making based on intuitionistic trapezoidal fuzzy number. Appl Math Model 37(5):3467–3477
Zheng G, Jing Y, Huang H, Gao Y (2010) Application of improved grey relational projection method to evaluate sustainable building envelope performance. Appl Energy 87(2):710–720

Part III
FMEA Based on Compromise Ranking MCDM Methods

Chapter 7
FMEA Using Fuzzy VIKOR Method

Fuzzy set theory is a way of addressing vague concepts, which provides a means for representing uncertainty involved in the real situation. On the other side, the VIKOR method is a recently developed multi-criteria decision-making (MCDM) method, which focuses on ranking and selecting from a set of alternatives in the presence of conflicting criteria and on proposing compromise solution(s). In (Liu et al. 2012), linguistic terms, expressed in trapezoidal or triangular fuzzy numbers, were used to assess the ratings and weights for risk factors. For selecting the most serious failure modes, an extended VIKOR method was used to determine the risk priorities of the failure modes that have been identified. As a result, a fuzzy FMEA model based on fuzzy set theory and VIKOR method was proposed for the prioritization of failure modes, specifically intended to address some limitations of the traditional FMEA.

7.1 Fuzzy Set Theory and VIKOR Method

7.1.1 Fuzzy Set Theory

Fuzzy set theory was developed by Zadeh (1965) to solve fuzzy phenomenon problems existing in the real world, such as uncertain, imprecise, unspecific, and fuzzy situations. This theory has an advantage over the traditional set theory when measuring the ambiguity of concepts that are associated with human beings' subjective judgments.

Definition 7.1 Let X be the universe of discourse, $X = \{x_1, x_2, \ldots, x_n\}$, a fuzzy set $\tilde{A}$ of X is characterized by a membership function $\mu_{\tilde{A}}(x)$, which associates with each element x in X a real number in the interval [0, 1]. The function value $\mu_{\tilde{A}}(x)$ is termed the grade of membership of x in $\tilde{A}$ (Zadeh 1965). The larger the $\mu_{\tilde{A}}(x)$ is, the stronger the grade of membership for x in $\tilde{A}$.

H.-C. Liu, *FMEA Using Uncertainty Theories and MCDM Methods*,
DOI 10.1007/978-981-10-1466-6_7

Definition 7.2 A fuzzy number is a fuzzy subset in the universe of discourse X whose membership function is both convex and normal (Chen 2001). A fuzzy set $\tilde{A}$ of the universe of discourse X is convex if and only if for all x_1, x_2 in X, $\mu_{\tilde{A}}(\lambda x_1 + (1-\lambda)x_2) \geq \min(\mu_{\tilde{A}}(x_1), \mu_{\tilde{A}}(x_2))$, where $\lambda \in [0,1]$. A fuzzy set $\tilde{A}$ of the universe of discourse X is called a normal fuzzy set implying that $\exists x_i \in X, \mu_{\tilde{A}}(x_i) = 1$.

Triangular and trapezoidal fuzzy numbers are the most common used fuzzy numbers in both theory and practice. In fact, triangular fuzzy numbers are special cases of trapezoidal fuzzy numbers. For the sake of simplicity and without loss of generality, trapezoidal fuzzy numbers are preferred for representing the linguistic variables in this chapter.

Definition 7.3 A positive trapezoidal fuzzy number $\tilde{A}$ can be denoted as (a_1, a_2, a_3, a_4), shown in Fig. 7.1. The membership function $\mu_{\tilde{A}}(x)$ is defined as:

$$\mu_{\tilde{A}}(x) = \begin{cases} 0, & x < a_1, \\ \frac{x-a_1}{a_2-a_1}, & a_1 \leq x \leq a_2, \\ 1, & a_2 \leq x \leq a_3, \\ \frac{x-a_4}{a_3-a_4}, & a_3 \leq x \leq a_4, \\ 0, & x > a_4. \end{cases} \tag{7.1}$$

where $[a_2, a_3]$ is called a mode interval of $\tilde{A}$, and a_1 and a_4 are called lower and upper limits of $\tilde{A}$, respectively.

Definition 7.4 Give any two positive trapezoidal fuzzy numbers $\tilde{A} = (a_1, a_2, a_3, a_4)$, $\tilde{B} = (b_1, b_2, b_3, b_4)$, and a positive real number r, the algebraic operations of trapezoidal fuzzy numbers are displayed as follows (Liu et al. 2012, 2014):

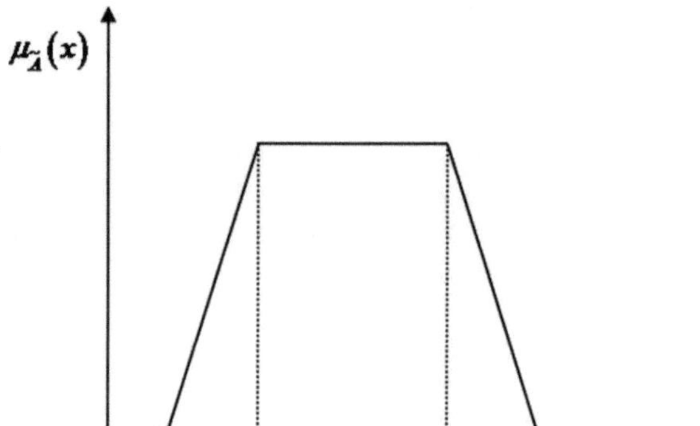

Fig. 7.1 Trapezoidal fuzzy number $\tilde{A}$

(1) $\tilde{A} \oplus \tilde{B} = [a_1 + b_1, a_2 + b_2, a_3 + b_3, a_4 + b_4]$,
(2) $\tilde{A} \ominus \tilde{B} = [a_1 - b_4, a_2 - b_3, a_3 - b_2, a_4 - b_1]$,
(3) $\tilde{A} \otimes \tilde{B} \cong [a_1 b_1, a_2 b_2, a_3 b_3, a_4 b_4]$,
(4) $\tilde{A} \oslash \tilde{B} \cong [a_1/b_4, a_2/b_3, a_3/b_2, a_4/b_1]$,
(5) $\tilde{A} \otimes r \cong [a_1 r, a_2 r, a_3 r, a_4 r]$.

Definition 7.5 A linguistic variable is a variable whose values are expressed in linguistic terms. The concept of linguistic variable is very useful in dealing with situations which are too complex or too ill-defined to be reasonably described by traditional quantitative expressions (Zadeh 1975).

The linguistic values can be represented by fuzzy numbers. In this chapter, the importance weights of risk factors and the fuzzy ratings of failure modes with respect to each risk factor are considered as linguistic variables. For example, these linguistic terms can be expressed in positive trapezoidal fuzzy numbers as Tables 7.1 and 7.2. Figures 7.2 and 7.3 show their membership functions for the sake of visualization. It should be noticed that the membership function values can be determined according to the historical data and the detailed questionnaire answered by domain experts (Liu et al. 2011).

An important step in fuzzy multi-criteria decision making is the defuzzification step which transforms a fuzzy number into a crisp value. Many different techniques for this transformation can be utilized, but the most commonly used defuzzification method is the centroid method, also known as the center of gravity (COG) or center of area (COA) defuzzification.

Table 7.1 Linguistic terms for rating failure modes

Linguistic terms	Fuzzy numbers
Very low (VL)	(0, 0, 1, 2)
Low (L)	(1, 2, 2, 3)
Medium low (ML)	(2, 3, 4, 5)
Medium (M)	(4, 5, 5, 6)
Medium high (MH)	(5, 6, 7, 8)
High (H)	(7, 8, 8, 9)
Very high (VH)	(8, 9, 10, 10)

Table 7.2 Linguistic terms for rating risk factor weights

Linguistic terms	Fuzzy numbers
Very low (VL)	(0, 0, 0.1, 0.2)
Low (L)	(0.1, 0.2, 0.2, 0.3)
Medium low (ML)	(0.2, 0.3, 0.4, 0.5)
Medium (M)	(0.4, 0.5, 0.5, 0.6)
Medium high (MH)	(0.5, 0.6, 0.7, 0.8)
High (H)	(0.7, 0.8, 0.8, 0.9)
Very high (VH)	(0.8, 0.9, 1, 1)

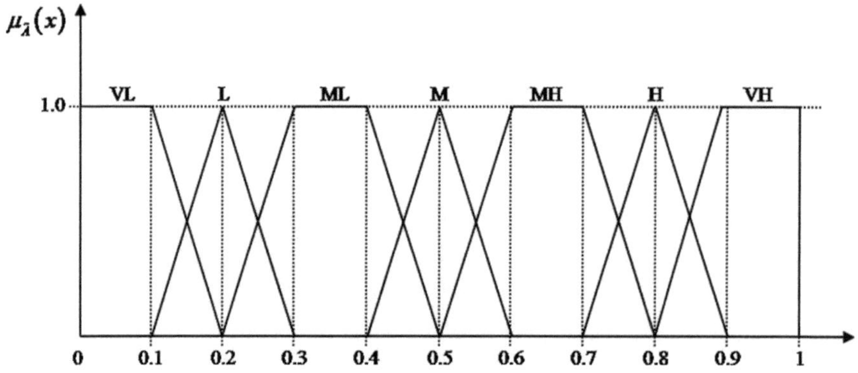

Fig. 7.2 Membership functions for rating risk factor weights (Liu et al. 2012)

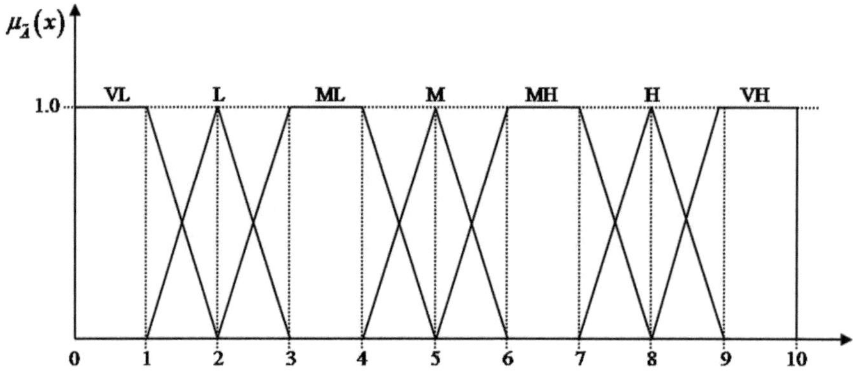

Fig. 7.3 Membership functions for rating failure modes (Liu et al. 2012)

Definition 7.6 The centroid defuzzification method can be expressed by the following relation:

$$\bar{x}(\tilde{A}) = \frac{\int x\mu_{\tilde{A}}(x)\mathrm{d}x}{\int \mu_{\tilde{A}}(x)\mathrm{d}x}, \tag{7.2}$$

where $\bar{x}(\tilde{A})$ is the defuzzified value. For a trapezoidal fuzzy number (a_1, a_2, a_3, a_4), the centroid-based defuzzified value turns out to be (Liu et al. 2012, 2014):

$$\bar{x}\left(\tilde{A}\right) = \frac{1}{3}\left[a_1 + a_2 + a_3 + a_4 - \frac{a_4a_3 - a_1a_2}{(a_4 + a_3) - (a_1 + a_2)}\right]. \tag{7.3}$$

7.1.2 The VIKOR Method

The VIKOR method was proposed by Opricovic and Tzeng (2002) for multi-criteria optimization of complex systems with the Serbian name: VlseKriterijumska Optimizacija I Kompromisno Resenje (means multi-criteria optimization and compromise solution) (Opricovic and Tzeng 2004). This method focuses on ranking and selecting from a set of alternatives and determines compromise solutions for a problem with conflicting criteria, which can help decision makers to reach a final decision. Here, the compromise solution is a feasible solution which is the closest to the ideal, and a compromise means an agreement established by mutual concessions (Opricovic and Tzeng 2007).

The VIKOR method introduces the multi-criteria ranking index based on the particular measure of closeness to the ideal solution (Opricovic 1998). This ranking index is an aggregation of all criteria, the relative importance of the criteria, and a balance between total and individual satisfaction. According to (Opricovic and Tzeng 2004), the multi-criteria measure for compromise ranking is developed from the L_p-metric utilized as an aggregating function in a compromise programming method. Suppose a set of m alternatives denoted as A_1, A_2, …, A_m. For an alternative A_i, the rating of the jth aspect is denoted by f_{ij}; i.e., f_{ij} is the value of jth criterion function for the alternative A_i; n is the number of criteria. Development of the VIKOR method started with the following form of L_p-metric:

$$L_{p,i} = \left\{ \sum_{j=1}^{n} \left[\frac{w_j \left(f_j^* - f_{ij} \right)}{f_j^* - f_j^-} \right]^p \right\}^{1/p}, \quad 1 \leq p \leq \infty,\ i = 1, 2, \ldots, m. \tag{7.4}$$

In the VIKOR method, $L_{1,i}$ (as S_i in Eq. 7.9) and $L_{\infty,i}$ (as R_i in Eq. 7.10) are used to formulate ranking measurements. The solution gained by min S_i is with a maximum group utility ("majority" rule), and the solution gained by min R_i is with a minimum individual regret of the "opponent."

Assume that a group MCDM problem has l decision makers DM_k $(k = 1, 2, \ldots, l)$, m alternatives A_i $(i = 1, 2, \ldots, m)$, and n decision criteria C_j $(j = 1, 2, \ldots, n)$. Each alternative is assessed with respect to the n criteria. All the performance ratings assigned to the alternatives with respect to each criterion form a decision matrix denoted by $X = \left[x_{ij}\right]_{m \times n}$. Then, the VIKOR method can be summarized as the following steps (Opricovic and Tzeng 2004, 2007; Liu et al. 2012):

Step 1. Pull the decision makers' opinions to get the aggregated fuzzy weights of criteria and construct a fuzzy decision matrix

Let the fuzzy rating and importance weight given by the kth decision maker be $\tilde{x}_{ij}^k = \left(x_{ij1}^k, x_{ij2}^k, x_{ij3}^k, x_{ij4}^k \right)$ and $\tilde{w}_j^k = \left(w_{j1}^k, w_{j2}^k, w_{j3}^k, w_{j4}^k \right)$, i = 1, 2, …, m and j = 1, 2, …, n, respectively. Hence, the aggregated fuzzy rating ($\tilde{x}_{ij}$) of alternatives with respect to each criterion is calculated as:

$$\tilde{x}_{ij} = \left(x_{ij1}, x_{ij2}, x_{ij3}, x_{ij4}\right) \tag{7.5}$$

where $x_{ij1} = \min\limits_k x_{ij1}^k, x_{ij2} = \frac{1}{l}\sum_{k=1}^{l} x_{ij2}^k, x_{ij3} = \frac{1}{l}\sum_{k=1}^{l} x_{ij3}^k, x_{ij4} = \max\limits_k x_{ij4}^k$.

The aggregated fuzzy weight ($\tilde{w}_j$) of each criterion is calculated as:

$$\tilde{w}_j = \left(w_{j1}, w_{j2}, w_{j3}, w_{j4}\right) \tag{7.6}$$

where $w_{j1} = \min\limits_k w_{j1}^k, w_{j2} = \frac{1}{l}\sum_{k=1}^{l} w_{j2}^k, w_{j3} = \frac{1}{l}\sum_{k=1}^{l} w_{j3}^k, w_{j4} = \max\limits_k w_{j4}^k$.

A MCDM problem can be concisely expressed in a matrix format as follows:

$$\tilde{D} = \begin{bmatrix} \tilde{x}_{11} & \tilde{x}_{12} & \ldots & \tilde{x}_{1n} \\ \tilde{x}_{21} & \tilde{x}_{22} & \ldots & \tilde{x}_{2n} \\ \vdots & \vdots & \ldots & \vdots \\ \tilde{x}_{m1} & \tilde{x}_{m2} & \ldots & \tilde{x}_{mn} \end{bmatrix}, \quad \tilde{W} = [\,\tilde{w}_1 \quad \tilde{w}_2 \quad \ldots \quad \tilde{w}_n\,],$$

where $\tilde{x}_{ij}$ is the rating of alternative A_i with respect to criterion C_j, $\tilde{w}_j$ is the importance weight of the jth criterion, and $\tilde{x}_{ij} = \left(x_{ij1}, x_{ij2}, x_{ij3}, x_{ij4}\right)$ and $\tilde{w}_j = \left(w_{j1}, w_{j2}, w_{j3}, w_{j4}\right)$ are linguistic terms which can be approximated by positive trapezoidal fuzzy numbers.

Step 2. Defuzzify the fuzzy decision matrix and fuzzy criteria weights into crisp values

The defuzzification of fuzzy decision matrix and the fuzzy weight of each criterion is done by using the centroid defuzzification method (Eq. 7.3).

Step 3. Determine the best f_j^* and the worst f_j^- values of all criteria ratings, $j = 1, 2, \ldots, n$

$$f_j^* = \begin{Bmatrix} \max\limits_i x_{ij}, & \text{for benefit criteria} \\ \min\limits_i x_{ij}, & \text{for cost criteria} \end{Bmatrix}, \quad j = 1, 2, \ldots, n, \tag{7.7}$$

$$f_j^- = \begin{Bmatrix} \min\limits_i x_{ij}, & \text{for benefit criteria} \\ \max\limits_i x_{ij}, & \text{for cost criteria} \end{Bmatrix}, \quad j = 1, 2, \ldots, n. \tag{7.8}$$

Step 4. Compute the values S_i and R_i, $i = 1, 2, \ldots, m$, by the relations

$$S_i = \sum_{j=1}^{n} \frac{w_j\left(f_j^* - x_{ij}\right)}{f_j^* - f_j^-}, \tag{7.9}$$

$$R_i = \max_j \left(\frac{w_j \left(f_j^* - x_{ij} \right)}{f_j^* - f_j^-} \right). \tag{7.10}$$

where w_j are the weights of criteria, expressing their relative importance.

Step 5. Compute the values Q_i, $i = 1, 2, \ldots, m$, by the relation

$$Q_i = v \frac{S_i - S^*}{S^- - S^*} + (1 - v) \frac{R_i - R^*}{R^- - R^*}, \tag{7.11}$$

where $S^* = \min_i S_i, S^- = \max_i S_i$, $R^* = \min_i R_i, R^- = \max_i R_i$ and v is introduced as a weight for the strategy of maximum group utility, whereas $1–v$ is the weight of the individual regret. Normally, the value of v can be set to 0.5.

Step 6. Rank the alternatives, sorting by the values S, R, and Q in increasing order. The results are three ranking lists.

Step 7. Propose a compromise solution, the alternative ($A^{(1)}$), which is the best ranked by the measure Q (minimum) if the following two conditions are satisfied:

C1. Acceptable advantage: $Q(A^{(2)}) - Q(A^{(1)}) \geq \text{DQ}$, where $A^{(2)}$ is the alternative with second position in the ranking list by Q; $\text{DQ} = 1/(m - 1)$.

C2. Acceptable stability in decision making: The alternative $A^{(1)}$ must also be the best ranked by S or/and R. This compromise solution is stable within a decision-making process, which could be: "voting by majority rule" (when $v > 0.5$ is needed), or "by consensus" $v \approx 0.5$, or "with veto" ($v < 0.5$).

If one of the conditions is not satisfied, then a set of compromise solutions is proposed, which consists of:

- Alternatives $A^{(1)}$ and $A^{(2)}$ if only the condition C2 is not satisfied or
- Alternatives $A^{(1)}$, $A^{(2)}$, …, $A^{(M)}$ if the condition C1 is not satisfied; $A^{(M)}$ is determined by the relation $Q(A^{(M)}) - Q(A^{(1)}) < \text{DQ}$ for maximum M (the positions of these alternatives are "in closeness").

7.2 The Proposed FMEA Model

It has been extensively argued that the risk factors O, S, and D are not easy to be precisely evaluated and the traditional FMEA takes no account of the relative importance of risk factors (Liu et al. 2012, 2013). Fuzzy logic is the tool for transforming the vagueness of human feeling and recognition and its decision-making ability into a mathematical formula. It also provides meaningful representation of measurement for uncertainties and vague concepts expressed in natural language. So, fuzzy MCDM methods are preferred instead of crisp decision-making methods for overcoming the limitations of the traditional FMEA procedure.

In this chapter, the risk factors and their relative importance weights are considered as linguistic variables. Because linguistic assessments merely approximate the subjective judgments of decision makers, we can consider linear trapezoidal membership functions to be adequate for capturing the vagueness of these linguistic assessments. A systematic approach to apply the VIKOR is proposed to determine the risk priorities of failure modes under a fuzzy environment in this section. The flowchart in Fig. 7.4 shows the proposed approach to rank the failure modes, which are identified in FMEA process.

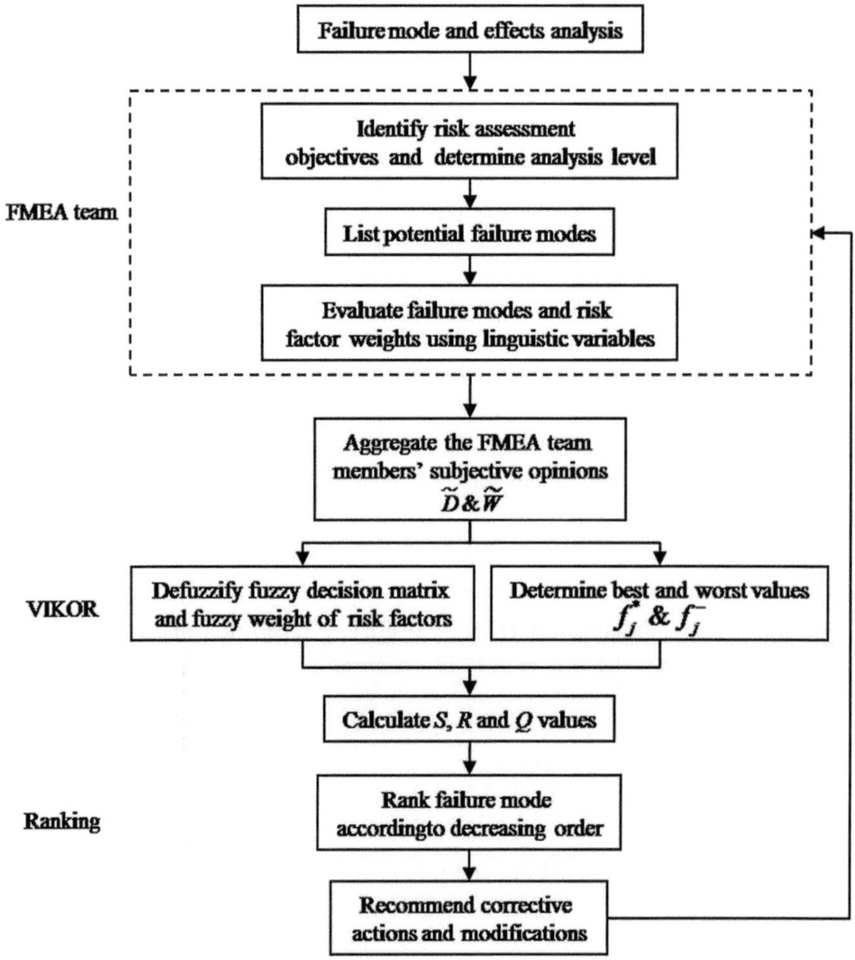

Fig. 7.4 Flowchart of the proposed FMEA approach (Liu et al. 2012)

To sum up, the risk priorities of failure modes are determined through the following steps (Liu et al. 2012):

Step 1. Identify the objectives of risk assessment and determine the risk analysis level.

Step 2. Arrange a FMEA team, list the potential failure modes, and describe a finite set of relevant risk factors.

Step 3. Determine appropriate linguistic variables for risk factors and their relative importance weights.

Step 4. Evaluate the importance of risk factors and the ratings of failure modes with respect to each risk factor using the linguistic variables.

Step 5. Apply the VIKOR approach:

- FMEA team members' linguistic evaluations regarding failure modes with respect to every risk factor and the risk factor weights are aggregated.
- Fuzzy decision matrix and fuzzy weight of each risk factor are defuzzified into crisp values.
- The best f_j^* and worst f_j^- values are determined.
- The values S, R, and Q are calculated, respectively.

Step 6. Determine the ranking orders of all failure modes according to the decreasing order of the values S, R, and Q.

Step 7. Analyze the results and develop recommendations to enhance the system performance.

7.3 An Illustrative Example

7.3.1 Application

The proposed FMEA model has been applied to the medical risk management of a tertiary care university hospital which aimed at preventing from medical accident by reducing medical mistakes and non-iatrogenic diseases. The risk analysis was carried out between August and October 2012, collecting data from doctors, anesthetists, and nurses working in the university hospital which had approximately 850 beds. This hospital was acknowledged as a typical hospital in the big urban areas of China. We analyze the risk of general anesthesia process because of its higher level of risk. The specific topic of the FMEA was determined through conversations among the department of anesthesiology, the department of surgery, and the hospital's quality management group.

In what follows, the steps of the risk assessment process are introduced in detail (Liu et al. 2012):

Step 1. The hospital desires to identify the most serious failure modes during the general anesthesia process to take appropriate measures in advance and prevent the incidence of medical errors.

Step 2. A FMEA team of five decision makers, DM1, DM2, DM3, DM4, and DM5, has been formed in the hospital in order to evaluate the identified failure modes. The team was composed of two anesthetists from the anesthesiology department, two chief physicians from the surgery department, and one operating room nursing supervisor. During the course of the FMEA, the FMEA team gathered information from interviews, meetings, and published materials. All potential failure modes of the general anesthesia process were identified by a session of systematic brainstorming. The FMEA director created a preliminary flow diagram after initial discussions by the FMEA team. The expert group then expanded and edited the diagram into its final form using their knowledge of various facets of the process. Given its complexity and length, the original flow diagram was divided into 15 subprocesses. To limit the FMEA to a more manageable scope, the team chose 6 of the 31 potential failure modes for the current investigation that were thought to have the system errors most in need of correction. The selected failure modes are arterial gas bolt (FM1), go esophageal (FM2), respiratory depression (FM3), not estimate surgery enough (FM4), blood transfusion wrong (FM5), and visceral injury (FM6).

Step 3. The third step of the proposed model is to define the risk factors, O, S, and D, using linguistics terms. In this regard, each variable is defined using membership functions that cover the universe of discourse of each variable. To define the linguistic terms for each variable, several interviews were arranged with the FMEA team members. The objective of the first meeting was to introduce our proposed risk priority model and to explore the options to implement this technique. Fortunately, the design of the risk matrix, as implemented by the hospital, is based on linguistic definitions for the three risk factors *O*, *S*, and *D*. Table 7.3 presents seven linguistic terms and their definition for the three risk factors. The next step entailed the development of the membership functions for each of the three risk factors. During this process, the FMEA team members were asked to define seven membership functions for each of the risk factors according to the definitions shown in Table 7.3. Triangular and trapezoidal membership function shapes were chosen, since they are intuitive to experts. To elicit the membership functions from the experts, questions similar to this were asked: "What is the degree of membership of 1 in 'Medium Low'?" Figure 7.3 shows the findings from the elicitation process. The membership functions for *O*, *S*, and *D* are similar as they both cover the same universe of discourse.

Step 4. The five FMEA team members use the linguistic weighting variables shown in Fig. 7.2 to assess the relative importance of risk factors. The importance weights of the risk factors determined by these five FMEA team members are shown in Table 7.4. Also the FMEA team members use the linguistic rating variables shown in Fig. 7.3 to evaluate the ratings of failure modes with respect to each risk factor. The ratings of the six failure modes by the FMEA team under the three risk factors are shown in Table 7.5.

Step 5-1. The linguistic evaluations shown in Tables 7.4 and 7.5 are converted into trapezoidal fuzzy numbers. Then the aggregated weights of risk factors and aggregated fuzzy ratings of failure modes are calculated to determine the fuzzy weight of each risk factor and construct the fuzzy decision matrix, as in Table 7.6.

Step 5-2. The crisp values for decision matrix and the weight of each risk factor are computed as shown in Table 7.7.

Step 5-3. In this step, the best and the worst values of all risk factor ratings are determined as follows:

$$f_O^* = 3.800, \quad f_S^* = 4.038, \quad f_D^* = 2.189,$$
$$f_O^- = 8.044, \quad f_S^- = 8.000, \quad f_D^- = 5.962.$$

Step 5-4. The values of S, R, and Q are calculated for all the failure modes as Table 7.8.

Step 6. The rankings of the failure modes by S, R, and Q in decreasing order are shown in Table 7.9.

As shown in Table 7.9, FM3 is apparently the most serious failure mode according to Q values and should be given the top risk priority by the hospital, and this will be followed by FM6, FM2, FM5, FM1, and FM4.

7.3.2 Comparisons and Discussion

In order to evaluate the proposed FMEA approach, we used the above case study to analyze some comparable methods, which include the conventional RPN method and the fuzzy TOPSIS method (Kutlu and Ekmekçioğlu 2012). Table 7.10 exhibits the ranking results of all the six failure modes as obtained using these approaches.

First, from Table 7.10, we can find that except for FM6, the risk priority orders of other failure modes obtained by the proposed method are all different from those by the conventional RPN method. The main reasons for these differences can be explained by the shortcomings of the traditional FMEA mentioned in Chap. 1. According to the conventional RPN method, FM5 (RPN = 192) is assumed to be more important and has a higher priority than FM3 (RPN = 162). However, the result of our proposed method shows that FM3 has a higher priority compared with

Table 7.3 Linguistic definitions of risk factors

Linguistic terms	*O*	*S*	*D*
Very low (VL)	No known occurrence	No injury to the patient or impact on the system	Error will always be detected
Low (L)	Rare failures (yearly)	Very minor to the patient	Very high probability that error will be detected
Medium low (ML)	Occasional failures (quarterly)	Minor injury to the patient	High probability of detection
Medium (M)	Monthly	Moderate injury to the patient	Moderate chance that error will be detected
Medium high (MH)	Frequent (weekly)	Moderate high injury to the patient	Remote chance of detection only
high (H)	Inevitable and predictable failure	May result in major injury to the patient	Remote or low likelihood of detection
Very high (VH)	Daily or every time	May cause death of the patient	No chance that error will be detected; no mechanism exists

Table 7.4 Importance weight of risk factors from the FMEA team (Liu et al. 2012)

Risk factors	Team members				
	DM1	DM2	DM3	DM4	DM5
O	H	H	VH	H	MH
S	VH	VH	H	VH	VH
D	MH	MH	M	H	MH

FM5, which tallies with the actual situation because the former has a higher severity rating and is therefore ranked higher than the latter. In addition, in the case of more than one failure modes having the same RPN values, such as FM1 and FM4, the proposed method can distinguish them from each other, thus providing more information than that of the traditional FMEA does. As expected, the proposed method considering the important weights of risk factors achieves a more accurate risk priority ranking, discriminating among the results far more accurate than the conventional RPN method.

Second, there is not much difference between the two sets of risk priority rankings obtained by the proposed method and the fuzzy TOPSIS method. But the most and least serious failure modes are the same, which are FM3 and FM4, respectively. The fuzzy TOPSIS method introduces two "reference" points, but it does not consider the relative importance of the distances from these points. Thus, the ranking produced by the fuzzy TOPSIS method may be biased. For example, based on the fuzzy TOPSIS method, FM6 is ranked behind FM2. However, in reality, the former is more important than the later and the results of the proposed method also show that FM6 has a higher priority in comparison with FM2. This is

Table 7.5 Judgments on failure modes by the FMEA team (Liu et al. 2012)

Risk factors		*O*					*S*					*D*				
Team members		DM1	DM2	DM3	DM4	DM5	DM1	DM2	DM3	DM4	DM5	DM1	DM2	DM3	DM4	DM5
Failure modes	FM1	M	M	M	MH	M	ML	ML	ML	M	M	M	ML	ML	ML	ML
	FM2	H	MH	H	MH	MH	H	MH	H	H	H	M	M	ML	M	M
	FM3	VH	MH	VH	VH	VH	MH	MH	MH	MH	MH	MH	M	MH	MH	M
	FM4	M	M	L	M	M	M	M	ML	M	M	VL	ML	VL	ML	VL
	FM5	M	ML	M	M	M	M	MH	MH	M	M	L	ML	L	L	L
	FM6	MH	H	M	MH	M	H	H	H	H	H	L	M	L	L	VL

Table 7.6 Aggregated fuzzy rating of failure modes and aggregated fuzzy weights of risk factors (Liu et al. 2012)

Failure modes	*O*	*S*	*D*
FM1	(4, 5.2, 5.4, 8)	(2, 3.8, 4.4, 6)	(2, 3.4, 4.2, 6)
FM2	(5, 6.8, 7.4, 9)	(5, 7.6, 7.8, 9)	(2, 4.6, 4.8, 6)
FM3	(5, 8.4, 9.4, 10)	(5, 6, 7, 8)	(4, 5.6, 6.2, 8)
FM4	(1, 4.4, 4.4, 6)	(2, 4.6, 4.8, 6)	(0, 1.2, 2.2, 5)
FM5	(2, 4.6, 4.8, 6)	(4, 5.4, 5.8, 8)	(1, 2.2, 2.4, 5)
FM6	(4, 6, 6.4, 9)	(7, 8, 8, 9)	(0, 2.2, 2.4, 6)
Weights	(0.5, 0.78, 0.82, 1)	(0.7, 0.88, 0.96, 1)	(0.4, 0.62, 0.68, 0.9)

Table 7.7 Crisp values for decision matrix and risk factor weights (Liu et al. 2012)

Failure modes	*O*	*S*	*D*
FM1	5.756	4.038	3.922
FM2	7.038	7.244	4.244
FM3	8.044	6.500	5.962
FM4	3.800	4.244	2.189
FM5	4.244	5.855	2.756
FM6	6.393	8.000	2.759
Weights	0.768	0.878	0.650

Table 7.8 The values of *S*, *R*, and *Q* for all the failure modes (Liu et al. 2012)

	Failure modes					
	FM1	FM2	FM3	FM4	FM5	FM6
S	0.653	1.650	1.964	0.046	0.581	1.445
R	0.354	0.710	0.768	0.046	0.403	0.878
Q	0.343	0.817	0.934	0	0.354	0.865

Table 7.9 The ranking of failure modes by *S*, *R*, and *Q* (Liu et al. 2012)

	Failure modes					
	FM1	FM2	FM3	FM4	FM5	FM6
By *S*	4	2	1	6	5	3
By *R*	5	3	2	6	4	1
By *Q*	5	3	1	6	4	2

also true for FM1 and FM5. Therefore, a more accurate ranking can be achieved by using the proposed fuzzy VIKOR method to evaluate the risk priority orders for failure analysis problems.

Table 7.10 Ranking comparisons

Failure modes	Proposed method	RPN method	Fuzzy TOPSIS
FM1	5	4	4
FM2	3	6	2
FM3	1	3	1
FM4	6	4	6
FM5	4	1	5
FM6	2	2	3

The analysis of the results produced by the fuzzy VIKOR, the conventional RPN, and the fuzzy TOPSIS methods shows that a more accurate, reasonable risk assessment can be obtained by applying the combination of fuzzy logic and VIKOR method.

7.3.3 Model Verification

To verify the validity of the proposed fuzzy FMEA model for risk management, a meeting was conducted with a group of six experts at the participating hospital consisting of two risk analysts, the manager of quality management office, two chief physicians, and the operating room nursing supervisor. The purpose of the meeting was to present the traditional approach of applying FMEA, its drawbacks, and the proposed risk priority model to address these limitations. Thereafter, the expert group was encouraged to raise questions and provide feedback. The feedback received on the proposed approach from all experts was positive. According to the domain experts, the proposed risk priority model is more suitable for the risk evaluation problem examined and can find the most critical failures effectively. Consequently, the reliability of the general anesthesia process can be assured by using the proposed risk assessment methodology.

Risk evaluation in FMEA is often influenced by uncertainty in real-life applications, and in such situation fuzzy set theory is an appropriate tool to deal with this kind of problems. In real decision-making process, the decision maker in FMEA team is unable (or unwilling) to express his assessments precisely in numerical values and the evaluations are very often expressed in linguistic terms. In this chapter, an extension of VIKOR, a recently introduced MCDM method, in fuzzy environment was used to deal with the risk factors and identify the most serious failure modes for corrective actions. The VIKOR method focuses on ranking and selecting from a set of alternatives in the presence of conflicting criteria. It determines a compromise solution that could be accepted by the decision makers. The provided case study has demonstrated the capability of the proposed FMEA model to deal with the risk evaluation problems in FMEA and to manage a criticality analysis in an intuitive and easy manner.

References

Chen CT (2001) A fuzzy approach to select the location of the distribution center. Fuzzy Sets Syst 118(1):65–73

Kutlu AC, Ekmekçioğlu M (2012) Fuzzy failure modes and effects analysis by using fuzzy TOPSIS-based fuzzy AHP. Expert Syst Appl 39(1):61–67

Liu HC, Liu L, Bian QH, Lin QL, Dong N, Xu PC (2011) Failure mode and effects analysis using fuzzy evidential reasoning approach and grey theory. Expert Syst Appl 38(4):4403–4415

Liu HC, Liu L, Liu N, Mao LX (2012) Risk evaluation in failure mode and effects analysis with extended VIKOR method under fuzzy environment. Expert Syst Appl 39(17):12926–12934

Liu HC, Liu L, Liu N (2013) Risk evaluation approaches in failure mode and effects analysis: a literature review. Expert Syst Appl 40(2):828–838

Liu HC, Fan XJ, Li P, Chen YZ (2014) Evaluating the risk of failure modes with extended MULTIMOORA method under fuzzy environment. Eng Appl Artif Intell 34:168–177

Opricovic S (1998) Multi-criteria optimization of civil engineering systems. Faculty of Civil Engineering, Belgrade

Opricovic S, Tzeng GH (2002) Multicriteria planning of post-earthquake sustainable reconstruction. Computer-Aided Civ Infrastruct Eng 17(3):211–220

Opricovic S, Tzeng GH (2004) Compromise solution by MCDM methods: a comparative analysis of VIKOR and TOPSIS. Eur J Oper Res 156(2):445–455

Opricovic S, Tzeng GH (2007) Extended VIKOR method in comparison with outranking methods. Eur J Oper Res 178(2):514–529

Zadeh LA (1965) Fuzzy sets. Inf Control 8(3):338–353

Zadeh LA (1975) The concept of a linguistic variable and its application to approximate reasoning–I. Inf Sci 8(3):199–249

Chapter 8
FMEA Using Intuitionistic Fuzzy Hybrid TOPSIS Approach

In Chap. 3, the theory of intuitionistic fuzzy sets (IFSs) has been proven to be useful for FMEA to deal with the vagueness and uncertainty existed in the risk-evaluating process. The technique for order preference by similarity to ideal solution (TOPSIS), proposed by Hwang and Yoon (1981), is one of the well-known MCDM methods and has been extensively applied to various engineering and management fields. In response, Liu et al. (2015) extended the classical TOPSIS method to the intuitionistic fuzzy environment and introduced an intuitionistic fuzzy hybrid TOPSIS (IFH-TOPSIS) approach to determine the risk priorities of failure modes in FMEA. Moreover, both subjective and objective weights of risk factors are taken into consideration in the process of risk and failure analysis. The proposed approach aims to represent a comprehensive criticality analysis methodology to overcome the limitations and improve the effectiveness of the traditional FMEA.

8.1 Preliminaries

Please refer to Sect. 3.1 for the basic definitions and operations related to IFSs, such as IFNs, the IFWA operator, and the IFHWED operator, which will be used in the proposed model.

8.2 The Proposed FMEA Approach

Usually, the risk factors O, S, and D are assessed with exact numbers in the traditional FMEA. However, in the real-life application, due to the increasing complexity of the assessed systems and the lack of knowledge or data about the problem domain, the risk factors are not easy to be precisely evaluated. As such, in

H.-C. Liu, *FMEA Using Uncertainty Theories and MCDM Methods*,
DOI 10.1007/978-981-10-1466-6_8

Table 8.1 Linguistic terms for rating failure modes

Linguistic terms	IFNs
Extremely low (EL)	(0.00, 0.90)
Very low (VL)	(0.10, 0.75)
Low (L)	(0.25, 0.60)
Medium low (ML)	(0.40, 0.50)
Medium (M)	(0.50, 0.45)
Medium high (MH)	(0.60, 0.25)
High (H)	(0.75, 0.10)
Very high (VH)	(0.90, 0.10)
Extremely high (EH)	(1.00, 0.00)

Table 8.2 Linguistic terms for rating risk factor weights

Linguistic terms	IFNs
Very low (VL)	(0.10, 0.85)
Low (L)	(0.30, 0.65)
Moderate (M)	(0.50, 0.50)
High (H)	(0.75, 0.20)
Very high (VH)	(0.90, 0.05)

this chapter, we choose linguistic terms for the assessment of risk factors and the individual evaluation grade is defined as an IFN. Table 8.1 shows the linguistic terms and their IFNs used for evaluating risk factors. Furthermore, both subjective and objective weights of risk factors are considered in the proposed FMEA. The subjective weights are assessed by decision makers or domain experts using the linguistic terms as provided in Table 8.2. The objective weights are determined by the ordered weights of the risk factors, which can be derived by the normal distribution-based method (Xu 2005).

The flowchart in Fig. 8.1 shows the proposed approach to rank the failure modes, which are identified in FMEA process. Three key steps are included in this approach, i.e., aggregation, calculation, and ranking. The FMEA team gives their individual judgments on failure modes by using linguistic terms defined by IFNs. The IFWA operator is cited for aggregating these judgments in order to form a consensus group judgment. Incorporated with the subjective and objective risk factor weights, the IFH-TOPSIS is used for calculating the relative closeness coefficients of the identified failure modes. Finally, the priority ranking of failure modes can be determined according to the results obtained in the previous step.

Suppose there are l cross-functional members $\mathrm{TM}_k\,(k = 1, 2, \ldots, l)$ in a FMEA team responsible for the assessment of m failure modes $\mathrm{FM}_i\,(i = 1, 2, \ldots, m)$ with respect to n risk factors $\mathrm{RF}_j\,(j = 1, 2, \ldots, n)$. Each team member TM_k is given a weight $\lambda_k > 0\,(k = 1, 2, \ldots, l)$ satisfying $\Sigma_{k=1}^{l}\lambda_k = 1$ to reflect his/her relative importance in the FMEA team. Let $\alpha_{ij}^k = \left(\mu_{ij}^k, v_{ij}^k\right)$ be the IFN provided by TM_k on the assessment of FM_i with respect to RF_j, and let $w_j^k = \left(\mu_j^k, v_j^k\right)$ be the subjective

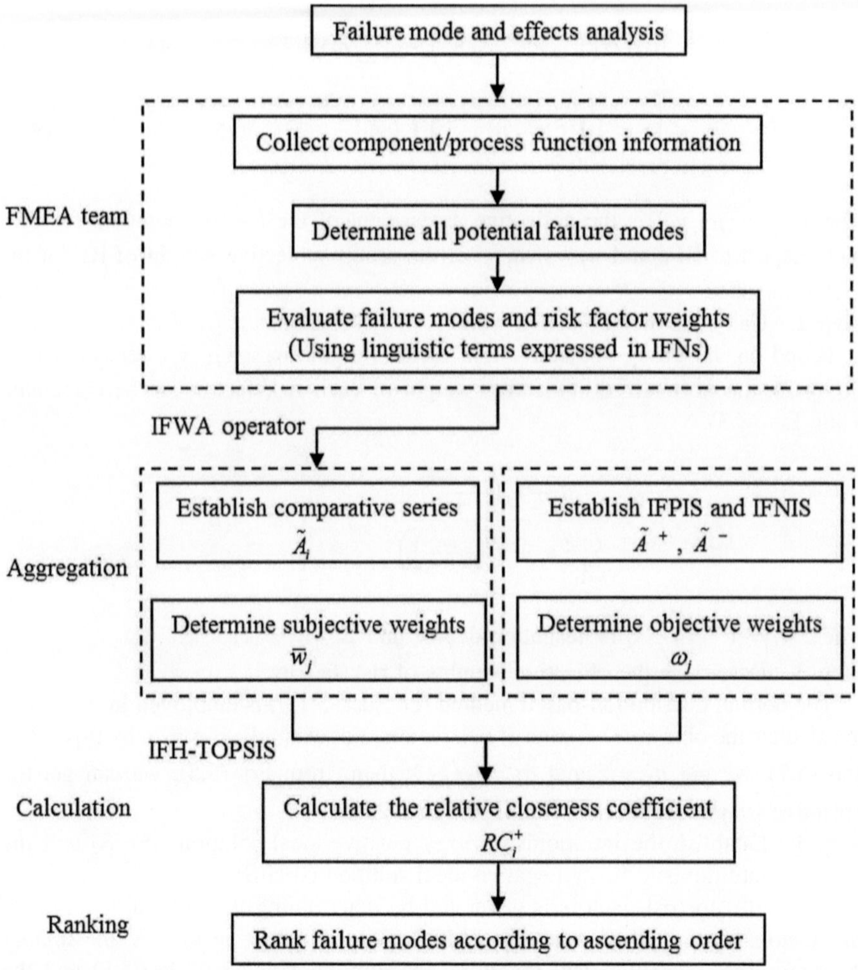

Fig. 8.1 Flowchart of the proposed FMEA approach (Liu et al. 2015)

weight of risk factor RF_j given by TM_k. Based on these assumptions, the m failure modes can be prioritized by employing the following steps (Liu et al. 2015):

Step 1. Aggregate the FMEA team members' subjective opinions by using the IFWA operator:

$$\alpha_{ij} = \text{IFWA}\left(\alpha_{ij}^1, \alpha_{ij}^2, \ldots, \alpha_{ij}^l\right) = \sum_{k=1}^{l} \lambda_k \alpha_{ij}^k,$$

$$= \left[1 - \prod_{k=1}^{l} \left(1 - \mu_{ij}^k\right)^{\lambda_k}, \prod_{k=1}^{l} \left(v_{ij}^k\right)^{\lambda_k}\right] \quad i = 1, 2, \ldots, m; j = 1, 2, \ldots, n, \quad (8.1)$$

$$w_j = \text{IFWA}\left(w_j^1, w_j^2, \ldots, w_j^l\right) = \sum_{k=1}^{l} \lambda_k w_j^k$$
$$= \left[1 - \prod_{k=1}^{l}\left(1 - \mu_j^k\right)^{\lambda_k}, \prod_{k=1}^{l}\left(v_j^k\right)^{\lambda_k}\right] \quad j = 1, 2, \ldots, n, \qquad (8.2)$$

where $\alpha_{ij} = (\mu_{ij}, v_{ij})$ is the collective assessment of the l team members for FM_i with respect to RF_j, and $w_j = (\mu_j, v_j)$ is the group subjective weight of RF_j of the l team members.

Step 2. Calculate the subjective weights of risk factors

Based on the group subjective risk factor weights $w_j = (\mu_j, v_j)$ determined by Eq. (8.2), the normalized subjective weight of each risk factor can be calculated using Eq. (8.3).

$$\bar{w}_j = \frac{\mu_j + \pi_j\left(\frac{\mu_j}{\mu_j + v_j}\right)}{\sum_{j=1}^{n}\left(\mu_j + \pi_j\left(\frac{\mu_j}{\mu_j + v_j}\right)\right)}, \quad j = 1, 2, \ldots, n, \qquad (8.3)$$

where $\pi_j = 1 - \mu_j - v_j$ is hesitation degree and $\Sigma_{j=1}^{n} \bar{w}_j = 1$.

Step 3. Determine the objective weights of risk factors

The normal distribution-based method (cf. Sect. 3.1.3) is employed in this paper to calculate the objective weights of risk factors. For example, if $n = 3$, by Eqs. (3.6) and (3.7), we get $\mu_3 = 2$ and $\sigma_3 = \sqrt{2/3}$; then, from Eq. (3.5), we can get the objective weight vector $\omega = (0.243, 0.514, 0.243)^T$.

Step 4. Establish the intuitionistic fuzzy positive-ideal solution (IFPIS) and the intuitionistic fuzzy negative-ideal solution (IFNIS)

The IFPIS for risk factors is generated by determining the optimal level of all risk factors for the failure modes in FMEA. When conducting FMEA, the smaller the score, the lesser the risk; therefore, the minimum value $\alpha^- = (0, 1)$ and the maximum value $\alpha^+ = (1, 0)$ can be used as the IFPIS and the IFNIS, respectively. They can be expressed as follows:

$$\tilde{A}^+ = \left[\alpha_1^+, \alpha_2^+, \ldots, \alpha_n^+\right] = \left[\alpha^-, \alpha^-, \ldots, \alpha^-\right], \qquad (8.4)$$

$$\tilde{A}^- = \left[\alpha_1^-, \alpha_2^-, \ldots, \alpha_n^-\right] = \left[\alpha^+, \alpha^+, \ldots, \alpha^+\right]. \qquad (8.5)$$

Step 5. Calculate the distances of each failure mode from the IFPIS and the IFNIS by using the IFHWED operator

A comparative series with n components or risk factors can be expressed as $\tilde{A}_i = \left[\alpha_{i1}, \alpha_{i2}, \ldots, \alpha_{in}\right]$, where α_{ij} $(j = 1, 2, \ldots, n)$ are obtained by Eq. (8.1). Then, the intuitionistic fuzzy hybrid weighted Euclidean distances, D_i^+ and D_i^-, of each failure mode from the positive-ideal and the negative-ideal solutions are calculated as follows:

$$D_i^+ = IFHWED(\tilde{A}_i, \tilde{A}^+)$$
$$= \varphi \sqrt{\sum_{j=1}^{n} \bar{w}_j \left(d_{IFD}\left(\alpha_{ij}, \alpha_j^+\right) \right)^2} + (1-\varphi) \sqrt{\sum_{j=1}^{n} \omega_j \left(d_{IFD}\left(\alpha_{i\sigma(j)}, \alpha_{\sigma(j)}^+\right) \right)^2},$$
$$i = 1, 2, \ldots m, \tag{8.6}$$

$$D_i^- = IFHWED(\tilde{A}_i, \tilde{A}^-)$$
$$= \varphi \sqrt{\sum_{j=1}^{n} \bar{w}_j \left(d_{IFD}\left(\alpha_{ij}, \alpha_j^-\right) \right)^2} + (1-\varphi) \sqrt{\sum_{j=1}^{n} \omega_j \left(d_{IFD}\left(\alpha_{i\sigma(j)}, \alpha_{\sigma(j)}^-\right) \right)^2}, \tag{8.7}$$
$$i = 1, 2, \ldots m,$$

where $(\sigma(1), \sigma(2), \ldots, \sigma(n))$ is any permutation of $(1, 2, \ldots, n)$, such that $d_{IFD}\left(\alpha_{i\sigma(j-1)}, \alpha_{\sigma(j-1)}^+\right) \geq d_{IFD}\left(\alpha_{i\sigma(j)}, \alpha_{\sigma(j)}^+\right)$, $j = 1, 2, \ldots, n$ and $d_{IFD}\left(\alpha_{i\sigma(j-1)}, \alpha_{\sigma(j-1)}^-\right) \geq d_{IFD}\left(\alpha_{i\sigma(j)}, \alpha_{\sigma(j)}^-\right)$, $j = 1, 2, \ldots, n$, and $\varphi \in [0, 1]$.

Step 6. Calculate the relative closeness coefficient of each failure mode to the IFPIS

The relative closeness coefficient of failure mode $\tilde{A}_i$ with respect to the IFPIS, $\tilde{A}^+$, is defined as follows:

$$RC_i^+ = \frac{D_i^-}{D_i^+ + D_i^-}, \quad i = 1, 2, \ldots, m, \tag{8.8}$$

where $0 \leq RC_i^+ \leq 1$.

Step 7. Rank all the failure modes

The smaller the RC_i^+, the bigger the overall risk and the higher the risk priority. As a result, the ranking order of all the failure modes can be determined according to the ascending order of their relative closeness coefficients.

8.3 An Illustrative Example

8.3.1 *Implementation*

To demonstrate the applicability of the proposed approach, a real case study concerning a 1.8-in. color super-twisted nematic (CSTN) (Chang and Wen 2010; Liu et al. 2015) is solved by the proposed IFH-TOPSIS method. On the whole, CSTN liquid crystal display (LCD) and thin-film transistor (TFT) LCD are two main streams in LCD technologies. The CSTN has a lower cost than TFT LCD. Thus, for the cost concern, CSTN LCD has major applications in mobile phones and MP3/MP4.

A FMEA team consisting of four experts identifies sixteen potential failure modes in the CSTN and needs to prioritize them in terms of their risk factors such as *O*, *S*, and *D* so that high risky failure modes can be corrected with top priority. The FMEA of this CSTN is presented in Table 8.3. Due to the difficulty in precisely assessing the risk factors and their relative importance weights, the FMEA team members are assumed to evaluate them by employing the linguistic terms expressed in IFNs in Tables 8.1 and 8.2. Subsequently, evaluations of the four experts in linguistic terms for the sixteen failure modes on each risk factor, and the risk factor weights are obtained as expressed in Table 8.4. The four experts from different departments are assumed to be of different importance because of their different domain knowledge and expertise. To reflect their differences in performing FMEA, the four experts are assigned the following relative weights: 0.15, 0.20, 0.30, and 0.35.

After converting into corresponding IFNs, the FMEA team members' individual assessments are aggregated into group assessments by using Eqs. (8.1)–(8.2). The results so obtained are presented in Table 8.5. After the determination of the group weights of risk factors by utilizing Eq. (8.3), the subjective weight vector for the risk factors is obtained as $\bar{w} = (0.370, 0.508, 0.122)$. On the other side, the objective weight vector for the risk factors can be derived as $\omega = (0.243, 0.514, 0.243)^T$ by using the normal distribution-based method.

In this step, using the subjective and the objective weight vectors of the risk factors and the group assessments of the four experts, the distances from the IFPIS and the IFNIS are calculated using Eqs. (8.6) and (8.7), respectively, for each failure mode identified in the FMEA. The results of these calculations are summarized in Table 8.6. In this example, the parameter φ is assumed to be 0.5, and the IFPIS and IFNIS are set as follows:

$$\tilde{A}^+ = \left[\alpha_1^+, \alpha_2^+, \alpha_3^+\right] = [(0,1),(0,1),(0,1)],$$

$$\tilde{A}^- = \left[\alpha_1^-, \alpha_2^-, \alpha_3^-\right] = [(1,0),(1,0),(1,0)].$$

Substituting the distances of each failure mode from the IFPIS and the IFNIS into Eq. (8.8) will give the relative closeness coefficients to the IFPIS for all the failure modes. Finally, the risk priority ranking of the identified sixteen failure modes can be determined in accordance with the ascending order of their relative closeness coefficients. The results are given in Table 8.6. As we can see, FM10 is apparently the failure mode with the maximum overall risk and should be given the top risk priority, followed by FM13, FM12, FM8,..., FM6 and FM7.

Table 8.3 FMEA for the 1.8-in. CSTN (Chang and Wen 2010; Liu et al. 2015)

Item	Description	Failure mode	Failure effect	Failure cause	Existing process control
1	Dimension design	Incorrect mechanical design	Parts interfere with each other during the module assembly	Incorrect parts design due to the dimensions	Design the module by using 3D software
2	Assembling the display	Incorrect mechanical design	Difficult to assemble the display	Lower yield rate for the assembly	Set up the design rule for different component parts
3	Flex	Low yield ratio	Supplier can not submit the quantity	It is small for the trace line and close to the edge for the pad	Use the precision mold to cut the outline
4	Flex	It is the short of non-adhesive copper in the further	Supplier can not submit the quantity	The material is specific	Use general material
5	Lightguide performance	Poor lightguide performance	Low or uneven luminance	Poor lightguide pattern design	Lightguide pattern simulation before tool making
6	Lightguide performance	Poor lightguide performance	Low or uneven luminance	Incorrect structure design OD the area for light through to lightguide	Confirm the performance again after having the lightguide sample
7	Flex assembly	Performance of display not good	Poor brightness	Light bar leaves holder	Use adhesive and bezel to fix the light bar
8	Interface design of module	Display mirror	LCD display NG	IC pad design mirror	Double-check drawing
9	LCM audible noise	LCM will have regular noise occurring	Makes the user feel uncomfortable	Improper IC software setting	Use those ICs that have been qualified
10	LCM audible noise	LCM will have regular noise occurring	Makes the user feel uncomfortable	Mechanical design can not isolate from noise	Mechanical design must consider the "echo" effect
11	Cross talk	Poor performance	Black line and white line on the panel	ITO impedance too high	Removal and first inspection

(continued)

Table 8.3 (continued)

Item	Description	Failure mode	Failure effect	Failure cause	Existing process control
12	Cross talk	Poor performance	Black line and white line on the panel	Vth cannot meet IC Vop	Inspection of the electric station of LCD
13	Cross talk	Poor performance	Black line and white line on the panel	Bias label tolerance too large	Control current of module for the IC
14	Contrast ratio	Poor performance	Selected driving conditions not sufficient to drive LCD to optimal display conditions	LCD driving voltage too high	Product engineer calculates correct driving voltage based on the driving conditions provided by the customer
15	Background color	Uneven background color	Uneven LCD background color (under lit-up backlight conditions)	Uneven cell gap	Use bonding seal application for negative and STN products
16	Background color	Uneven background color	Uneven LCD background color (under lit-up backlight conditions)	Uneven cell gap	Full electrical testing

8.3.2 *Sensitivity Analysis*

A sensitivity analysis by changing the weights of the FMEA team members is calculated according to the information given in Table 8.7. For example, Case 0 shows the original weight values of the four experts, while the other cases show different weight values for possible situations. The ranking results of the sixteen failure modes for the considered cases are represented in Fig. 8.2. It is clearly shown in Fig. 8.2 that different combinations of experts' weights have great influence on the final ranking results in the failure analysis. Thus, proper determination of the relative weights of experts plays an essential role in the FMEA process. In general, the weights of FMEA team members can be determined by using direct rating, point allocation, eigenvector method, or Delphi method, etc., together with their domain knowledge (Chin et al. 2009; Liu et al. 2015). If there is no sufficient reason or evidence to show the differences among FMEA team members in their judgment qualities, the team members can be assigned an equal weight.

Table 8.4 Assessed information on the sixteen failure modes by the FMEA team (Liu et al. 2015)

Risk factors		*O*				*S*				*D*			
Team members		TM1	TM2	TM3	TM4	TM1	TM2	TM3	TM4	TM1	TM2	TM3	TM4
Weights		H	M	H	M	H	VH	H	VH	VL	L	L	VL
Failure modes	FM1	L	ML	L	ML	M	H	MH	ML	L	L	M	ML
	FM2	L	ML	ML	ML	M	M	M	ML	L	ML	ML	M
	FM3	ML	M	ML	ML	MH	M	MH	H	L	L	VL	EL
	FM4	ML	M	MH	M	M	M	ML	M	ML	L	ML	ML
	FM5	L	ML	ML	L	M	H	ML	M	ML	ML	EL	L
	FM6	L	ML	L	VL	M	H	ML	M	ML	L	ML	VL
	FM7	VL	VL	EL	L	MH	ML	H	MH	VL	L	L	EL
	FM8	L	VL	L	VL	H	VH	VH	EH	L	VL	VL	ML
	FM9	M	M	ML	M	M	MH	ML	M	L	EL	VL	L
	FM10	H	MH	ML	MH	ML	M	MH	MH	ML	M	MH	M
	FM11	VL	L	ML	L	M	MH	MH	M	L	ML	L	ML
	FM12	ML	ML	ML	ML	MH	H	MH	M	ML	ML	L	ML
	FM13	VL	L	L	ML	MH	H	MH	MH	M	M	MH	M
	FM14	ML	ML	L	L	H	M	ML	M	VL	L	ML	VL
	FM15	ML	M	ML	ML	M	MH	M	M	EL	VL	VL	EL
	FM16	M	ML	L	ML	M	ML	M	ML	ML	ML	L	ML

Table 8.5 Aggregated assessments on failure modes and the subjective weights of risk factors (Liu et al. 2015)

Failure modes	O	S	D
FM1	(0.337, 0.543)	(0.566, 0.290)	(0.386, 0.516)
FM2	(0.380, 0.514)	(0.467, 0.467)	(0.418, 0.495)
FM3	(0.421, 0.490)	(0.645, 0.204)	(0.124, 0.739)
FM4	(0.519, 0.383)	(0.472, 0.464)	(0.373, 0.519)
FM5	(0.329, 0.548)	(0.540, 0.344)	(0.244, 0.636)
FM6	(0.235, 0.626)	(0.540, 0.344)	(0.277, 0.598)
FM7	(0.129, 0.733)	(0.623, 0.218)	(0.148, 0.715)
FM8	(0.171, 0.678)	(1.000, 0.000)	(0.240, 0.629)
FM9	(0.472, 0.464)	(0.495, 0.413)	(0.161, 0.696)
FM10	(0.579, 0.268)	(0.556, 0.312)	(0.519, 0.383)
FM11	(0.279, 0.587)	(0.553, 0.335)	(0.337, 0.543)
FM12	(0.400, 0.500)	(0.606, 0.256)	(0.358, 0.528)
FM13	(0.287, 0.582)	(0.636, 0.208)	(0.532, 0.377)
FM14	(0.306, 0.563)	(0.524, 0.371)	(0.232, 0.635)
FM15	(0.421, 0.490)	(0.522, 0.400)	(0.051, 0.822)
FM16	(0.376, 0.520)	(0.447, 0.477)	(0.358, 0.528)
$\bar{w}$	(0.634, 0.331)	(0.849, 0.093)	(0.206, 0.743)

Table 8.6 Distance measures and relative closeness coefficients for the sixteen failure modes (Liu et al. 2015)

Failure modes	D^+	D^-	RC^+	Ranking
FM1	0.572	0.570	0.4991	7
FM2	0.510	0.576	0.5304	9
FM3	0.602	0.577	0.4894	5
FM4	0.553	0.534	0.4913	6
FM5	0.524	0.618	0.5412	11
FM6	0.505	0.651	0.5631	15
FM7	0.522	0.717	0.5787	16
FM8	0.666	0.626	0.4845	4
FM9	0.522	0.589	0.5302	8
FM10	0.690	0.444	0.3915	1
FM11	0.533	0.611	0.5341	10
FM12	0.598	0.540	0.4745	3
FM13	0.642	0.525	0.4499	2
FM14	0.506	0.636	0.5569	14
FM15	0.505	0.630	0.5551	13
FM16	0.495	0.602	0.5488	12

Table 8.7 Weights of FMEA team members regarding the considered cases (Liu et al. 2015)

Team members	Case 0	Case 1	Case 2	Case 3	Case 4
TM1	0.15	0.20	0.30	0.35	0.25
TM2	0.20	0.30	0.35	0.15	0.25
TM3	0.30	0.35	0.15	0.20	0.25
TM4	0.35	0.15	0.20	0.30	0.25

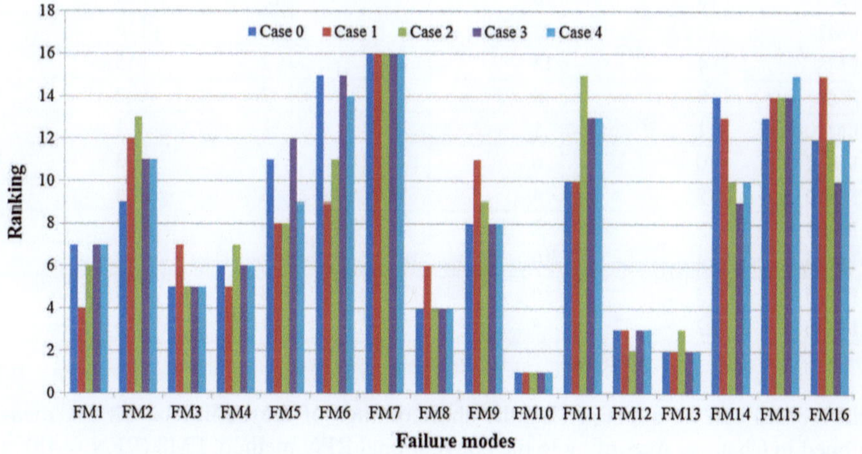

Fig. 8.2 Sensitivity analysis for the case study (Liu et al. 2015)

In addition, the two kinds of risk factor weights are assumed to be equally important and the parameter φ is taken as 0.5 in this case study. However, it can also be set according to actual situations and experts' opinions in real-life applications (Liu et al. 2014). Section 3.5.2 gives the main principles to determine the weight restriction φ.

8.3.3 Comparisons and Discussion

In order to evaluate the proposed FMEA, we used the above case study to analyze some comparable methods, which include the conventional RPN method, the intuitionistic fuzzy TOPSIS (IF-TOPSIS) (Boran et al. 2009), the fuzzy TOPSIS (Braglia et al. 2003), and the integrated weight-based fuzzy TOPSIS (IWF-TOPSIS) (Song et al. 2013). Table 8.8 exhibits the ranking results of all the identified failure modes as obtained using these approaches.

First, from Table 8.8, we can find that except for FM9, FM10, and FM13, the risk priority rankings of other failure modes obtained by the proposed method are different from those by the conventional RPN method. The main reasons for the

Table 8.8 Ranking comparisons (Liu et al. 2015)

Failure modes	Proposed method	RPN method	IF-TOPSIS	Fuzzy TOPSIS	IWF-TOPSIS
FM1	7	6	7	9	10
FM2	9	10	13	13	8
FM3	5	9	3	4	5
FM4	6	3	6	6	2
FM5	11	14	11	11	11
FM6	15	10	16	15	14
FM7	16	15	8	16	16
FM8	4	13	1	2	15
FM9	8	8	9	7	3
FM10	1	1	2	1	1
FM11	10	6	12	10	13
FM12	3	4	5	3	4
FM13	2	2	4	5	7
FM14	14	10	14	12	12
FM15	13	16	10	8	9
FM16	12	5	15	14	6

differences can be explained by the shortcomings of the traditional FMEA mentioned in Chap. 1. According to the conventional RPN method, FM8 (RPN = 48) is assumed to be more important and has a higher priority than FM3 (RPN = 32). However, the result of our proposed method shows that FM3 has a higher priority compared with FM8, which tallied with the actual situation because the former has a higher severity rating and is therefore ranked higher than the latter. In addition, in the case of more than one failure mode having the same RPN values, such as FM1 and FM11, and FM2, FM6, and FM14, the proposed method can distinguish them from each other, thus providing more information than that of the traditional FMEA.

Second, there are some differences between the two sets of risk priority rankings obtained by the proposed FMEA and the IF-TOPSIS. The ranking produced by the IF-TOPSIS method does not take into account the ordered weights of risk factors and thus results in biased conclusions. For example, based on the IF-TOPSIS method, FM2 is ranked behind FM15. However, in reality, the former is more important, and thus, the result of the proposed method shows that FM2 has a higher priority in comparison with FM15. This is also true for FM2 and FM6. In addition, FM8 turned out to be the most critical failure mode in terms of the IF-TOPSIS, while by using the proposed method, it ranks only on the fourth place and FM10 becomes the most critical one at the same time. Ranking FM10 as the first place can also be validated by the conventional RPN, the fuzzy TOPSIS, and the IWF-TOPSIS methods. As for FM8, it seems that there is no agreement on the results obtained by the five FMEA methods. A close look at the values of the risk factors for FM8 (O = 2, S = 8, D = 2) reveals that it has the highest value of S, which is the most important and is in line with the subjective risk factor weights. Thus, FM8 is given top priorities if only the subjective weights of risk factors are taken into account as in the IF-TOPSIS and the fuzzy TOPSIS, and it ranks only in the thirteenth or fifteenth position if the risk factors are assumed to have the same importance (as in the conventional RPN method), or both subjective and objective weights are considered (as in the IWF-TOPSIS). But the application of the proposed approach considering both the subjective and the objective weights of risk factors leads to FM8 to rank in the fourth priority position. This is mainly because in the proposed FMEA, the ordered weights for the risk factors are derived by the normal distribution-based method, which can relieve the influence of unduly high or unduly low values on the decision results by giving them low weights.

By that analogy, the main difference between the proposed FMEA and the fuzzy TOPSIS is that the subjective weights of risk factors are not considered during the fuzzy TOPSIS-based risk analysis, which may lead to unreasonable ranking of failure modes. Although both subjective and objective weights of risk factors are integrated in the IWF-TOPSIS, its objective weighting computation method is different from that of the proposed IFH-TOPSIS and Shannon entropy concept is used to calculate the objective risk factor weights. However, applying the entropy method is much more laborious and time-consuming. In addition, all fuzzy assessments provided by the FMEA team members must be converted into crisp values in the first step. This produces a loss of information and hence a lack of precision in the final results. Also, both the fuzzy TOPSIS and the IWF-TOPSIS are fuzzy logic-based FMEA methods, which may not be suitable in situations where precise membership functions are difficult or impossible to specify.

The analysis of the results produced by the traditional FMEA, the IF-TOPSIS, the fuzzy TOPSIS, the IWF-TOPSIS, and the IFH-TOPSIS methods shows that a more accurate and reasonable risk assessment can be derived by applying a combination of the IFHWED operator and the TOPSIS method. Compared with the convention RPN method and its various improvements, the FMEA model introduced in this chapter has the following properties: (1) It allows experts to evaluate the risk factors and their relative weights in linguistic variables which can be expressed by intuitionistic fuzzy sets. This added flexibility can extend the applicability of FMEA to systems where safety data are unavailable or unreliable. (2) Both subjective and objective weights of risk factors are taken into consideration in the process of prioritization of failure modes, which makes the proposed FMEA more realistic and more practical. (3) The proposed FMEA is not limited to the traditional three risk factors *O*, *S*, and *D*, but applicable to other kinds of risk factors (e.g., cost, time, and maintenance). That is, a potentially larger number of risk factors can be introduced if necessary. (4) The proposed model is simple and effective for failure mode risk evaluation, and in particular, the defined IFH-TOPSIS method offered a new way for prioritizing failure modes in FMEA.

References

Boran FE, Genç S, Kurt M, Akay D (2009) A multi-criteria intuitionistic fuzzy group decision making for supplier selection with TOPSIS method. Expert Syst Appl 36(8):11363–11368

Braglia M, Frosolini M, Montanari R (2003) Fuzzy TOPSIS approach for failure mode, effects and criticality analysis. Qual Reliab Eng Int 19(5):425–443

Chang KH, Wen TC (2010) A novel efficient approach for DFMEA combining 2-tuple and the OWA operator. Expert Syst Appl 37(3):2362–2370

Chin KS, Wang YM, Poon GKK, Yang JB (2009) Failure mode and effects analysis using a group-based evidential reasoning approach. Comput Oper Res 36(6):1768–1779

Hwang CL, Yoon K (1981) Multiple Attributes Decision Making Methods and Applications. Springer, Berlin

Liu HC, Liu L, Li P (2014) Failure mode and effects analysis using intuitionistic fuzzy hybrid weighted Euclidean distance operator. Int J Syst Sci 45(10):2012–2030

Liu HC, You JX, Shan MM, Shao LN (2015) Failure mode and effects analysis using intuitionistic fuzzy hybrid TOPSIS approach. Soft Comput 19(4):1085–1098

Song W, Ming X, Wu Z, Zhu B (2013) Failure modes and effects analysis using integrated weight-based fuzzy TOPSIS. Int J Comput Integr Manuf 26(12):1172–1186

Xu ZS (2005) An overview of methods for determining OWA weights. Int J Intell Syst 20(8):843–865

Part IV
FMEA Based on Other MCDM Methods

Chapter 9
FMEA Using Fuzzy DEMATEL Technique

The decision-making trial and evaluation laboratory (DEMATEL) technique is a comprehensive method that supports MCDM problems in building and analyzing a structural model involving causal relationships between components of a system. This method applies matrices and digraphs for visualizing the structure of complicated causal relationships and can identify key alternatives based on the type of relationships and degree of influences among them. Therefore, Liu et al. (2015) proposed a new risk assessment methodology based on fuzzy DEMATEL to rank the risk of failures in system FMEA. The new method can not only address some of the inherent limitations of the traditional FMEA but also cope with the interdependencies among various failures in fuzzy environment. As is illustrated by the numerical example, the proposed method is a suitable and effective method for prioritization of failures in system FMEA.

9.1 Fuzzy Sets and Fuzzy DEMATEL

9.1.1 Fuzzy Sets

The basic definitions of fuzzy sets (Zadeh 1975) can be found in Sect. 7.1.1, and in the following, we only introduce some basic concepts and operations which will be used in the proposed model.

Definition 9.1 A triangular fuzzy number $\tilde{a}$ can be denoted as (a_1, a_2, a_3), and its membership function $\mu_{\tilde{a}}(x)$ (as shown in Fig. 9.1) can be defined as follows:

H.-C. Liu, *FMEA Using Uncertainty Theories and MCDM Methods*,
DOI 10.1007/978-981-10-1466-6_9

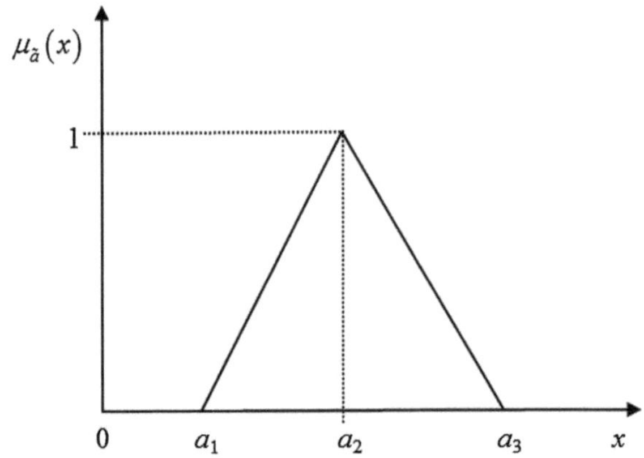

Fig. 9.1 Triangular fuzzy number $\tilde{a}$

$$\mu_{\tilde{a}}(x) = \begin{cases} 0, & x < a_1 \\ \frac{x-a_1}{a_2-a_1}, & a_1 \leq x \leq a_2 \\ \frac{a_3-x}{a_3-a_2}, & a_2 \leq x \leq a_3 \\ 0, & x > a_3 \end{cases}, \tag{9.1}$$

where $a_1 \leq a_2 \leq a_3$ and a_1, a_2, and a_3 denote the smallest possible value, the most promising value, and the largest possible value of a fuzzy event, respectively.

Definition 9.2 Given any two positive triangular fuzzy numbers $\tilde{a} = (a_1, a_2, a_3)$ and $\tilde{b} = (b_1, b_2, b_3)$ and a positive real number r, the algebraic operations of triangular fuzzy numbers can be expressed as follows (Liu et al. 2015):

(1) $\tilde{a} + \tilde{b} = [a_1 + b_1, a_2 + b_2, a_3 + b_3]$,
(2) $\tilde{a} - \tilde{b} = [a_1 - b_3, a_2 - b_2, a_3 - b_1]$,
(3) $\tilde{a} \times \tilde{b} \cong [a_1 b_1, a_2 b_2, a_3 b_3]$,
(4) $r \times \tilde{a} = [ra_1, ra_2, ra_3]$,
(5) $\tilde{a} \div \tilde{b} \cong [a_1/b_3, a_2/b_2, a_3/b_1]$,

Definition 9.3 According to (Liou and Wang 1992), the fuzzy weighted average (FWA) of n fuzzy numbers can be expressed as follows:

$$f(\tilde{a}_1, \tilde{a}_2, \ldots, \tilde{a}_n; \tilde{w}_1, \tilde{w}_2, \ldots, \tilde{w}_n) = \frac{\tilde{w}_1 \tilde{a}_1 + \tilde{w}_2 \tilde{a}_2 + \cdots + \tilde{w}_n \tilde{a}_n}{\tilde{w}_1 + \tilde{w}_2 + \cdots + \tilde{w}_n} = \frac{\sum_{i=1}^{n} \tilde{w}_i \tilde{a}_i}{\sum_{i=1}^{n} \tilde{w}_i}, \tag{9.2}$$

where $\tilde{a}_1, \tilde{a}_2, \ldots, \tilde{a}_n$ are the n positive fuzzy numbers to be weighted and $\tilde{w}_1, \tilde{w}_2, \ldots, \tilde{w}_n$ are their fuzzy weights.

Definition 9.4 The canonical representation of operation on triangular fuzzy numbers (Chou 2003), which is based on the graded mean integration representation method, is a popular defuzzification method converting fuzzy numbers into crisp scores. The graded mean integration representation of the triangular fuzzy number $\tilde{a} = (a_1, a_2, a_3)$ can be expressed by

$$\bar{x}_0(\tilde{a}) = \frac{1}{6}(a_1 + 4 \times a_2 + a_3), \tag{9.3}$$

where $\bar{x}_0(\tilde{a})$ is the defuzzified value.

9.1.2 *Fuzzy DEMATEL*

The DEMATEL method, presented by the Geneva Research Centre of the Battelle Memorial Institute (Gabus and Fontela 1973), is a comprehensive method for building and analyzing a structural model involving causal relationships between complex factors. It is especially practical and useful for visualizing the structure of complicated causal relationships with matrices or digraphs (Wu and Lee 2007). However, in real-life and real-world situations, the relationships between causes and effects are often complex and subtle. Moreover, the judgments and preferences of decision makers are often hard to quantify in exact numerical values due to the inherent vagueness of human language. Thus, fuzzy set theory was applied to DEMATEL for handling problems characterized by vagueness and imprecision.

Suppose a system contains a set of elements $S = \{E_1, E_2, \ldots, E_n\}$ and particular pairwise relations are determined for modeling with respect to a mathematical relation. The analytical procedure of the fuzzy DEMATEL method can be briefly described as follows (Lin and Wu 2004, 2008; Liu et al. 2015):

Step 1. Generate the initial direct-relation fuzzy matrix

$$\tilde{Z} = \begin{bmatrix} 0 & \tilde{z}_{12} & \cdots & \tilde{z}_{1n} \\ \tilde{z}_{21} & 0 & \cdots & \tilde{z}_{2n} \\ \vdots & \vdots & \ddots & \vdots \\ \tilde{z}_{n1} & \tilde{z}_{n2} & \cdots & 0 \end{bmatrix}, \tag{9.4}$$

where $\tilde{z}_{ij} = (z_{ij1}, z_{ij2}, z_{ij3})$ are triangular fuzzy numbers and $z_{ii}, i = 1, 2, \ldots, n$, will be regarded as the triangular fuzzy number $(0, 0, 0)$ whenever necessary.

Step 2. Normalize the initial direct-relation fuzzy matrix

$$\tilde{X} = \frac{\tilde{Z}}{r}, \tag{9.5}$$

where

$$\tilde{X} = \begin{bmatrix} \tilde{x}_{11} & \tilde{x}_{12} & \cdots & \tilde{x}_{1n} \\ \tilde{x}_{21} & \tilde{x}_{22} & \cdots & \tilde{x}_{2n} \\ \vdots & \vdots & \ddots & \vdots \\ \tilde{x}_{n1} & \tilde{x}_{n2} & \cdots & \tilde{x}_{nn} \end{bmatrix} \tag{9.6}$$

and

$$r = \max_{1 \le i \le n} \left(\sum_{j=1}^{n} z_{ij3} \right). \tag{9.7}$$

It is assumed at least one i such that $\sum_{j=1}^{n} z_{ij3} < r$.

Step 3. Obtain the total-relation fuzzy matrix

$$\tilde{T} = \lim_{k \to \infty} \left(\tilde{X}^1 + \tilde{X}^2 + \cdots + \tilde{X}^k \right) = \tilde{X}(1 - \tilde{X})^{-1}, \text{when} \lim_{k \to \infty} \tilde{X}^k = O. \tag{9.8}$$

Then,

$$\tilde{T} = \begin{bmatrix} \tilde{t}_{11} & \tilde{t}_{12} & \cdots & \tilde{t}_{1n} \\ \tilde{t}_{21} & \tilde{t}_{22} & \cdots & \tilde{t}_{2n} \\ \vdots & \vdots & \ddots & \vdots \\ \tilde{t}_{n1} & \tilde{t}_{n2} & \cdots & \tilde{t}_{nn} \end{bmatrix}, \tag{9.9}$$

where $\tilde{t}_{ij} = (t_{ij1}, t_{ij2}, t_{ij3})$ and

$$\begin{aligned} T_1 &= [t_{ij1}]_{n \times n} = X_1(I - X_1)^{-1}, \\ T_2 &= [t_{ij2}]_{n \times n} = X_2(I - X_2)^{-1}, \\ T_3 &= [t_{ij3}]_{n \times n} = X_3(I - X_3)^{-1}, \end{aligned} \tag{9.10}$$

in which $X_1 = [x_{ij1}]_{n \times n}, X_2 = [x_{ij2}]_{n \times n}, X_3 = [x_{ij3}]_{n \times n}$, and I is denoted as the identity matrix. The elements of the triangular fuzzy numbers in the total-relation fuzzy matrix $\tilde{T}$ are divided into T_1, T_2, and T_3, and $T_1 \prec T_2 \prec T_3$, when $x_{ij1} < x_{ij2} < x_{ij3}$ for any $i,j \in \{1, 2, \ldots, n\}$.

Step 4. Produce the causal diagram

By producing the total-relation fuzzy matrix $\tilde{T}$, then it is calculated $\tilde{R}_i + \tilde{C}_i$ and $\tilde{R}_i - \tilde{C}_i$ in which $\tilde{R}_i$ and $\tilde{C}_i$ are the sum of rows and the sum of columns of $\tilde{T}$, respectively. Next, the fuzzy numbers of $\tilde{R}_i + \tilde{C}_i$ and $\tilde{R}_i - \tilde{C}_i$ should be converted to crisp values by using Eq. (9.3). A causal diagram can be acquired by mapping the ordered pairs of $(\tilde{R}_i + \tilde{C}_i)^{\mathrm{def}}$ and $(\tilde{R}_i - \tilde{C}_i)^{\mathrm{def}}$, where the horizontal axis $(\tilde{R}_i + \tilde{C}_i)^{\mathrm{def}}$ is called "prominence" and the vertical axis $(\tilde{R}_i - \tilde{C}_i)^{\mathrm{def}}$ is called "relation." In the causal diagram, the prominence axis shows how much importance the element has, whereas the relation axis will divide the elements into cause and effect groups. In general, if the value $(\tilde{R}_i - \tilde{C}_i)^{\mathrm{def}}$ is positive, the element belongs to the cause group, and if the value $(\tilde{R}_i - \tilde{C}_i)^{\mathrm{def}}$ is negative, the element belongs to the effect group. Therefore, the causal diagram can visualize the complicated causal relationships between elements into a visible structural model and will provide valuable insights for decision making.

9.2 The Proposed FMEA Model

In this chapter, the importance weights of risk factors and the fuzzy ratings of failure modes with respect to each risk factor are considered as linguistic variables. For example, these linguistic variables can be expressed in triangular fuzzy numbers as shown in Tables 9.1 and 9.2. Then, the FWA is used to aggregate the FMEA team members' subjective opinions and to obtain the fuzzy risk priority number (FRPN) of each failure mode. In addition, as any failure mode may impact each other, the fuzzy DEMATEL technique is used to acquire the influenced structure between the failure modes in FMEA. After knowing the influenced structure among the failure modes, we rank them by the relations (direct/indirect)

Table 9.1 Linguistic terms for rating failure modes

Linguistic terms	Fuzzy numbers
Very low (VL)	(0, 0, 1)
Low (L)	(0, 1, 3)
Medium low (ML)	(1, 3, 5)
Medium (M)	(3, 5, 7)
Medium high (MH)	(5, 7, 9)
High (H)	(7, 9, 10)
Very high (VH)	(9, 10, 10)

Table 9.2 Linguistic terms for rating risk factor weights

Linguistic variables	Fuzzy numbers
Very low (VL)	(0, 0, 0.25)
Low (L)	(0, 0.25, 0.5)
Medium (M)	(0.25, 0.5, 0.75)
High (H)	(0.5, 0.75, 1)
Very high (VH)	(0.75, 1, 1)

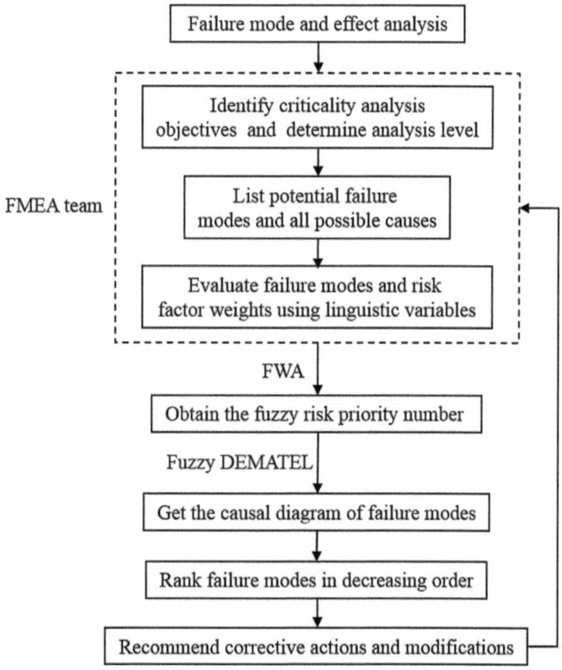

Fig. 9.2 Flowchart of the proposed fuzzy FMEA (Liu et al. 2015)

and prominences to find the most risk failure modes that will help to take relevant modifications. In short, the framework of risk evaluation contains three main phases: (1) obtaining the FRPN of each failure mode by the FWA, (2) constructing the causal diagram among failure modes by the fuzzy DEMATEL, and (3) ranking failure modes based on the relationships and influences between them. The flowchart in Fig. 9.2 shows the proposed methodology to prioritize the potential failure modes, which are identified in the FMEA process.

To sum up, the risk priorities of failure modes can be determined by employing the following steps (Liu et al. 2015):

Step 1. Identify the objectives of criticality analysis and determine the risk analysis level.

Step 2. Arrange the FMEA team, list potential failure modes and their causes, and describe a finite set of relevant risk factors.

Step 3. Determine the appropriate linguistic variables for risk factors and their relative importance weights.

Step 4. Evaluate the importance of the risk factors and the ratings of failure modes with respect to each risk factor using the linguistic variables.

Step 5. The FWA is used to obtain the FRPN of each failure mode.

Suppose there are l cross-functional members, $\mathrm{TM}_k(k = 1, 2, \ldots, l)$, in a FMEA team responsible for the assessment of m failure modes, $\mathrm{FM}_i(i = 1, 2, \ldots, m)$, with respect to n risk factors, $\mathrm{RF}_j(j = 1, 2, \ldots, n)$. Each team member TM_k is given a weight $\lambda_k > 0 (k = 1, 2, \ldots, l)$ satisfying $\Sigma_{k=1}^{l} \lambda_k = 1$ to reflect his/her relative importance in the FMEA process. Let $\tilde{x}_{ij}^k = \left(x_{ij1}^k, x_{ij2}^k, x_{ij3}^k\right)$ be the fuzzy rating provided by TM_k on the assessment of FM_i with respect to RF_j, and let $\tilde{w}_j^k = \left(w_{j1}^k, w_{j2}^k, w_{j3}^k\right)$ be the fuzzy weight of risk factor RF_j given by TM_k. Then, the FMEA team members' opinions can be aggregated by:

$$\tilde{x}_{ij} = \sum_{k=1}^{l} \lambda_k \tilde{x}_{ij}^k = \left(\sum_{k=1}^{l} \lambda_k x_{ij1}^k, \sum_{k=1}^{l} \lambda_k x_{ij2}^k, \sum_{k=1}^{l} \lambda_k x_{ij3}^k\right), \quad i = 1, 2, \ldots, m; j = 1, 2, \ldots, n, \tag{9.11}$$

$$\tilde{w}_j = \sum_{k=1}^{l} \lambda_k \tilde{w}_j^k = \left(\sum_{k=1}^{l} \lambda_k w_{j1}^k, \sum_{k=1}^{l} \lambda_k w_{j2}^k, \sum_{k=1}^{l} \lambda_k w_{j3}^k\right), \quad j = 1, 2, \ldots, n. \tag{9.12}$$

where $\tilde{x}_{ij}$ is the aggregated fuzzy rating of FM_i with respect to RF_j and $\tilde{w}_j$ is the aggregated fuzzy weight of RF_j.

Thus, the FRPN of each failure mode can be computed by using the FWA as below:

$$FRPN_i = \frac{\tilde{w}_1 \tilde{x}_{i1} + \tilde{w}_2 \tilde{x}_{i2} + \cdots + \tilde{w}_n \tilde{x}_{in}}{\tilde{w}_1 + \tilde{w}_2 + \cdots + \tilde{w}_n}, \quad i = 1, 2, \ldots, m. \tag{9.13}$$

Step 6. The fuzzy DEMATEL is utilized to get the causal diagram among failure modes.

- The initial direct-relation fuzzy matrix is generated.

The graph that can describe the interrelationships between the failure modes of a system is shown in Fig. 9.3, where nodes indicate the failures or causes of failures and directed connections (edges) indicate the effects of failures together. Also, the linguistic terms or triangular fuzzy numbers represent the degrees of direct influence among the failure modes. The initial direct-relation fuzzy matrix can be yielded by utilizing the FWA approach in terms of expert's knowledge as discussed in Step 5.

- The normalized direct-relation fuzzy matrix is acquired.
- The total-relation fuzzy matrix is calculated.
- The causal diagram is acquired, finally.

Step 7. Determine the ranking order of all failure modes according to the decreasing order of $(\tilde{R}_i + \tilde{C}_i)^{\text{def}}$ and $(\tilde{R}_i - \tilde{C}_i)^{\text{def}}$.

Step 8. Analyze the results and develop recommendations to enhance the system performance.

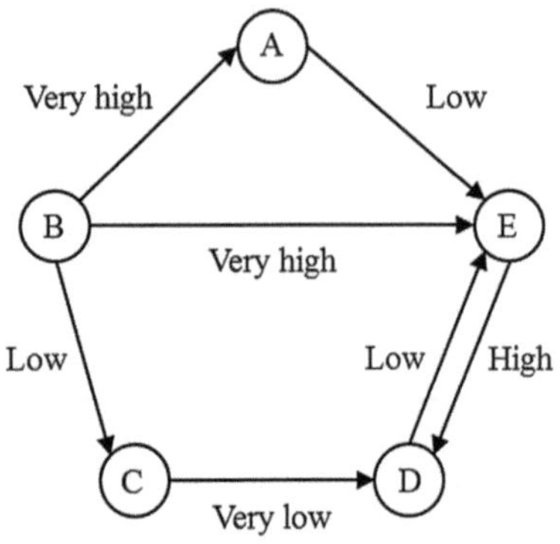

Fig. 9.3 An example of the directed graph

9.3 An Illustrative Example

9.3.1 Implementation

In this section, a real-world application in a thin-film transistor liquid crystal display (TFT-LCD) product (Chang and Cheng 2011; Liu et al. 2015) is employed to illustrate the capability of the proposed risk assessment methodology. The TFT-LCD was drawn from a professional liquid crystal display (LCD) manufacturer in Taiwan. The corporation's main products are small- to medium-sized LCD displays and modules, including TN, STN, CSTN, TFT, touch panel, and BLM, to wide applications for mobile phone, MP3, PDA, PMP, and DMP. The FMEA of the TFT-LCD product identified by a FMEA team is presented in Table 9.3.

Suppose a FMEA team is made up of five experts, each playing a different role in the team and given a different weight. The weights for the five members are assumed to be 0.25, 0.3, 0.2, 0.15, and 0.1, respectively. The FMEA team needs to evaluate the failure modes with respect to the three major risk factors, *O*, *S*, and *D*, so that high-risk failure modes can be corrected with top priority. Each team member evaluates the risk factors and their importance weights using the linguistic terms defined in Tables 9.1 and 9.2. The assessment information provided by the five team members is shown in Table 9.4.

Based on the information in Table 9.4, the five team members' subjective assessments are first aggregated by Eqs. (9.11) and (9.12). Then, the FPRNs of the identified failure modes are calculated by Eq. (9.13). The results so obtained are provided in Table 9.5. Since the risk ratings and the risk factor weights are all fuzzy numbers, the overall risk of each failure mode will be also a fuzzy number.

In addition, as shown in Table 9.5, the TFT-LCD product has 11 potential failure modes (FMs) and 15 causes of failure (CFs). Therefore, the corresponding directed graph consists of 26 nodes and 16 connections, as shown in Fig. 9.4. According to the results of Fig. 9.4, we can obtain the initial direct-relation fuzzy

Table 9.3 The FMEA of the TFT-LCD product (Liu et al. 2015)

No.	Potential failure modes	Causes of failure
1	Grayscale display defect (FM1)	Poor gamma curve design (CF1)
2	Uneven splotches at edges and corners of LCD (FM2)	Edge and interior delta and not the same (CF2)
3	Uneven splotches at edges and corners of LCD (FM2)	Silver paste and perimeter sealant material characteristics (CF3)
4	Uneven splotches at edges and corners of LCD (FM2)	Conductive material not able to cover CP dot area (CF4)
5	Flickering display (FM3)	Moisture seeps into Vcom CP dot and reduces conductivity (CF5)
6	Flickering display (FM3)	Liquid crystal resistance too low (CF6)
7	Flickering display (FM3)	Insufficient Cst capacitance setting (CF7)
8	No displays (FM4)	1. Short circuit by particle
		2. ITO scratch (CF8)
9	Missing pixels (FM5)	1. Etching failure
		2. Particle remains on LCD internal (CF9)
10	Missing lines (FM6)	1. Etching failure
		2. Particle remains on LCD internal (CF9)
11	Contrast ratio (FM7)	Poor operation by the operators (CF10)
12	Cross talk (FM8)	1. ITO impedance too high
		2. Vth cannot meet IC Vop
		3. Bias-level tolerance too large. (CF11)
13	Liquid crystal response time too slow (FM9)	1. Liquid crystal selection error
		2. Cell gap setting error (CF12)
14	Poor high-temperature contrast (FM10)	Liquid crystal clearing point too low (CF13)
15	Poor high-temperature contrast (FM10)	Spacer leaking light (CF14)
16	Bright region transmissiveness too low (FM11)	Poor liquid crystal Δn and LCD cell gap matching (CF15)

matrix $\tilde{Z}$ of the TFT-LCD product, as shown in Fig. 9.5. As Fig. 9.5 depicts, there is a unilateral flow only from CFs to FMs and the rest entries are equal to zero (three zero blocks). In other words, FMs have not any influence on CFs in this example. From matrix $\tilde{Z}$, the normalized direct-relation fuzzy matrix $\tilde{X}$ is calculated by using Eqs. (9.5)–(9.7). Finally, following Eqs. (9.8)–(9.10), the total-relation fuzzy matrix $\tilde{T}$ can be derived, as indicated in Fig. 9.6. Then, the values of $\tilde{R}_i, \tilde{C}_i, \tilde{R}_i + \tilde{C}_i$, and $\tilde{R}_i - \tilde{C}_i$ are obtained and given in Table 9.6. Afterward, fuzzy values of $\tilde{R}_i + \tilde{C}_i$ and $\tilde{R}_i - \tilde{C}_i$ are defuzzified by utilizing Eq. (9.3). The causal diagram could be generated by mapping the data set of $\left((\tilde{R}_i + \tilde{C}_i)^{\text{def}}, (\tilde{R}_i - \tilde{C}_i)^{\text{def}}\right)$. The outcomes resulted from the implementation of fuzzy DEMATEL for the TFT-LCD product are shown in Table 9.7 and Fig. 9.7.

Table 9.4 Assessed information on failure modes given by the FMEA team (Liu et al. 2015)

Risk factors		O					S					D				
Team members		TM1	TM2	TM3	TM4	TM5	TM1	TM2	TM3	TM4	TM5	TM1	TM2	TM3	TM4	TM5
No.	1	ML	ML	ML	L	L	MH	MH	M	MH	M	M	ML	ML	ML	ML
	2	ML	L	L	L	ML	M	M	M	M	M	ML	ML	ML	ML	ML
	3	ML	L	ML	L	ML	M	M	M	M	M	ML	L	ML	L	L
	4	ML	ML	ML	ML	ML	M	M	M	M	M	ML	L	L	L	ML
	5	M	ML	ML	ML	M	MH	M	M	MH	MH	L	L	L	ML	ML
	6	ML	ML	M	ML	ML	MH	M	M	MH	MH	M	ML	M	M	ML
	7	ML	L	ML	ML	L	MH	M	M	MH	MH	ML	ML	L	L	L
	8	L	L	ML	L	ML	H	MH	H	MH	H	ML	M	ML	ML	ML
	9	ML	ML	ML	ML	L	MH	MH	MH	MH	MH	M	M	MH	M	M
	10	ML	ML	ML	ML	L	MH	MH	MH	MH	MH	M	M	MH	M	M
	11	ML	ML	ML	ML	ML	M	M	MH	MH	MH	M	MH	M	ML	M
	12	ML	M	M	ML	M	ML	M	M	MH	M	ML	ML	M	ML	M
	13	M	ML	ML	M	ML	M	M	M	M	M	L	ML	ML	L	L
	14	L	ML	L	L	L	MH	M	M	MH	MH	ML	L	L	L	L
	15	ML	L	ML	L	L	MH	M	M	MH	MH	ML	ML	ML	ML	ML
	16	L	L	L	ML	ML	MH	MH	MH	M	M	ML	L	L	L	ML
Weights		M	M	L	L	M	VH	VH	H	VH	H	L	VL	L	VL	VL

Table 9.5 Aggregated assessment information and FPRNs for failure modes (Liu et al. 2015)

No.	Failure modes	Causes of failure	O	S	D	FRPN
1	FM1	CF1	(0.75, 2.5, 4.5)	(4.4, 6.4, 8.4)	(1.5, 3.5, 5.5)	(1.527, 5.066, 15.97)
2	FM2	CF2	(0.35, 1.7, 3.7)	(3, 5, 7)	(1, 3, 5)	(1.028, 3.906, 13.449)
3	FM2	CF3	(0.55, 2.1, 4.1)	(3, 5, 7)	(0.45, 1.9, 3.9)	(1.044, 3.934, 13.29)
4	FM2	CF4	(1, 3, 5)	(3, 5, 7)	(0.35, 1.7, 3.7)	(1.08, 4.175, 13.915)
5	FM3	CF5	(1.7, 3.7, 5.7)	(4, 6, 8)	(0.25, 1.5, 3.5)	(1.47, 4.997, 15.576)
6	FM3	CF6	(1.4, 3.4, 5.4)	(4, 6, 8)	(2.2, 4.2, 6.2)	(1.446, 5.121, 16.507)
7	FM3	CF7	(0.6, 2.2, 4.2)	(4, 6, 8)	(0.55, 2.1, 4.1)	(1.381, 4.616, 14.649)
8	FM4	CF8	(0.3, 1.6, 3.6)	(6.1, 8.1, 9.55)	(1.6, 3.6, 5.6)	(2.057, 5.902, 16.675)
9	FM5	CF9	(0.9, 2.8, 4.8)	(5, 7, 9)	(3.4, 5.4, 7.4)	(1.739, 5.681, 17.746)
10	FM6	CF9	(0.9, 2.8, 4.8)	(5, 7, 9)	(3.4, 5.4, 7.4)	(1.739, 5.681, 17.746)
11	FM7	CF10	(1, 3, 5)	(3.9, 5.9, 7.9)	(3.3, 5.3, 7.3)	(1.38, 5.028, 16.548)
12	FM8	CF11	(2.2, 4.2, 6.2)	(2.8, 4.8, 6.8)	(1.6, 3.6, 5.6)	(1.11, 4.536, 15.448)
13	FM9	CF12	(1.8, 3.8, 5.8)	(3, 5, 7)	(0.5, 2, 4)	(1.144, 4.426, 14.678)
14	FM10	CF13	(0.3, 1.6, 3.6)	(4, 6, 8)	(0.25, 1.5, 3.5)	(1.357, 4.399, 13.915)
15	FM10	CF14	(0.45, 1.9, 3.9)	(4, 6, 8)	(1, 3, 5)	(1.369, 4.601, 14.801)
16	FM11	CF15	(0.25, 1.5, 3.5)	(4.5, 6.5, 8.5)	(0.35, 1.7, 3.7)	(1.52, 4.705, 14.519)
Weights			(0.163, 0.413, 0.663)	(0.675, 0.925, 1)	(0, 0.113, 0.363)	

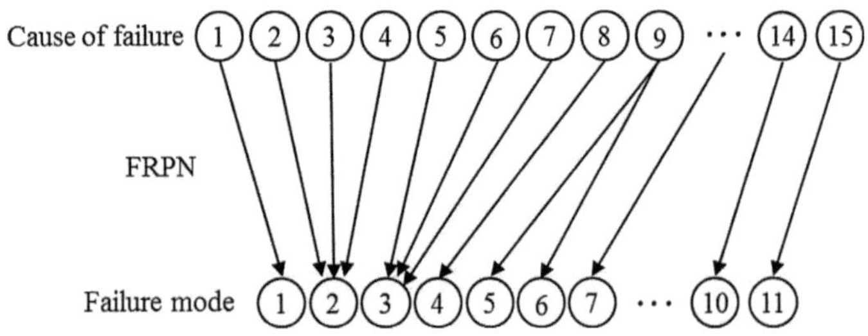

Fig. 9.4 Corresponding FMEA directed graph (Liu et al. 2015)

	CF1	CF2	CF3	...	CF14	CF15	FM1	FM2	FM3	...	FM10	FM11
CF1							(1.5, 5.1, 16.0)					
CF2								(1.0, 4.0, 13.5)				
CF3								(1.0, 3.9, 13.3)				
⋮			0				⋮	⋮	⋮	⋮	⋮	⋮
CF14											(1.4, 4.6, 14.8)	
CF15												(1.5,4.7, 14.5)
FM1												
FM2												
FM3			0							0		
⋮												
FM10												
FM11												

Fig. 9.5 The initial direct-relation fuzzy matrix $\tilde{Z}$ (Liu et al. 2015)

	CF1	CF2	CF3	...	CF14	CF15	FM1	FM2	FM3	...	FM10	FM11
CF1							(0.04,0.15,0.46)					
CF2								(0.03,0.11,0.39)				
CF3								(0.03,0.11,0.39)				
⋮			0				⋮	⋮	⋮	⋮	⋮	⋮
CF14											(0.04,0.13,0.43)	
CF15												(0.04,0.14,0.42)
FM1												
FM2												
FM3			0							0		
⋮												
FM10												
FM11												

Fig. 9.6 The total-relation fuzzy matrix $\tilde{T}$ (Liu et al. 2015)

From Table 9.7 and Fig. 9.7, it can be observed that CF9 apparently has the most influence on failure modes according to the value of $(\tilde{D}_i - \tilde{R}_i)^{\text{def}}$ and should be given the top risk priority by the corporation. The ranking order of CFs on the TFTLCD product that was obtained by combining the FWA and the fuzzy DEMATEL in terms of the risk factors is as follows: CF9 $\succ$ CF8 $\succ$ CF6 $\succ$ CF10 $\succ \cdots \succ$ CF2 $\succ$ CF3. Additionally, the causal diagram (Fig. 9.7) shows that FM3, with the largest $(\tilde{D}_i + \tilde{R}_i)^{\text{def}}$, is the most important failure mode for the TFTLCD product. Moreover, FM3, with the most negative value of $(\tilde{D}_i - \tilde{R}_i)^{\text{def}}$, is also the most easily improved one of all the failure modes;

Table 9.6 The values of $\tilde{R}_i, \tilde{C}_i, \tilde{R}_i + \tilde{C}_i$, and $\tilde{R}_i - \tilde{C}_i$ (Liu et al. 2015)

No.	$\tilde{R}_i$	$\tilde{C}_i$	$\tilde{R}_i + \tilde{C}_i$	$\tilde{R}_i - \tilde{C}_i$
CF1	(0.044, 0.146, 0.4604)	(0, 0, 0)	(0.044, 0.146, 0.4604)	(0.044, 0.146, 0.4604)
CF2	(0.0296, 0.1126, 0.3877)	(0, 0, 0)	(0.0296, 0.1126, 0.3877)	(0.0296, 0.1126, 0.3877)
CF3	(0.0301, 0.1134, 0.3831)	(0, 0, 0)	(0.0301, 0.1134, 0.3831)	(0.0301, 0.1134, 0.3831)
CF4	(0.0311, 0.1204, 0.4012)	(0, 0, 0)	(0.0311, 0.1204, 0.4012)	(0.0311, 0.1204, 0.4012)
CF5	(0.0424, 0.144, 0.4491)	(0, 0, 0)	(0.0424, 0.144, 0.4491)	(0.0424, 0.144, 0.4491)
CF6	(0.0417, 0.1476, 0.4759)	(0, 0, 0)	(0.0417, 0.1476, 0.4759)	(0.0417, 0.1476, 0.4759)
CF7	(0.0398, 0.1331, 0.4223)	(0, 0, 0)	(0.0398, 0.1331, 0.4223)	(0.0398, 0.1331, 0.4223)
CF8	(0.0593, 0.1701, 0.4807)	(0, 0, 0)	(0.0593, 0.1701, 0.4807)	(0.0593, 0.1701, 0.4807)
CF9	(0.0993, 0.3201, 1)	(0, 0, 0)	(0.0993, 0.3201, 1)	(0.0993, 0.3201, 1)
CF10	(0.0398, 0.145, 0.4771)	(0, 0, 0)	(0.0398, 0.145, 0.4771)	(0.0398, 0.145, 0.4771)
CF11	(0.032, 0.1308, 0.4454)	(0, 0, 0)	(0.032, 0.1308, 0.4454)	(0.032, 0.1308, 0.4454)
CF12	(0.033, 0.1276, 0.4231)	(0, 0, 0)	(0.033, 0.1276, 0.4231)	(0.033, 0.1276, 0.4231)
CF13	(0.0391, 0.1268, 0.4012)	(0, 0, 0)	(0.0391, 0.1268, 0.4012)	(0.0391, 0.1268, 0.4012)
CF14	(0.0395, 0.1326, 0.4267)	(0, 0, 0)	(0.0395, 0.1326, 0.4267)	(0.0395, 0.1326, 0.4267)
CF15	(0.0438, 0.1356, 0.4186)	(0, 0, 0)	(0.0438, 0.1356, 0.4186)	(0.0438, 0.1356, 0.4186)
FM1	(0, 0, 0)	(0.044, 0.146, 0.46)	(0.044, 0.146, 0.46)	(−0.044, −0.146, −0.46)
FM2	(0, 0, 0)	(0.091, 0.346, 1.172)	(0.091, 0.346, 1.172)	(−0.091, −0.346, −1.172)
FM3	(0, 0, 0)	(0.124, 0.425, 1.347)	(0.124, 0.425, 1.347)	(−0.124, −0.425, −1.347)
FM4	(0, 0, 0)	(0.059, 0.17, 0.481)	(0.059, 0.17, 0.481)	(−0.059, −0.17, −0.481)
FM5	(0, 0, 0)	(0.05, 0.164, 0.512)	(0.05, 0.164, 0.512)	(−0.05, −0.164, −0.512)
FM6	(0, 0, 0)	(0.049, 0.156, 0.488)	(0.049, 0.156, 0.488)	(−0.049, −0.156, −0.488)
FM7	(0, 0, 0)	(0.04, 0.145, 0.477)	(0.04, 0.145, 0.477)	(−0.04, −0.145, −0.477)
FM8	(0, 0, 0)	(0.032, 0.131, 0.445)	(0.032, 0.131, 0.445)	(−0.032, −0.131, −0.445)
FM9	(0, 0, 0)	(0.033, 0.128, 0.423)	(0.033, 0.128, 0.423)	(−0.033, −0.128, −0.423)
FM10	(0, 0, 0)	(0.079, 0.259, 0.828)	(0.079, 0.259, 0.828)	(−0.079, −0.259, −0.828)
FM11	(0, 0, 0)	(0.044, 0.136, 0.419)	(0.044, 0.136, 0.419)	(−0.044, −0.136, −0.419)

Table 9.7 The values of $(\tilde{R}_i + \tilde{C}_i)^{\text{def}}$ and $(\tilde{R}_i - \tilde{C}_i)^{\text{def}}$ and risk priority ranking (Liu et al. 2015)

No.	$(\tilde{R}_i + \tilde{C}_i)^{\text{def}}$	$(\tilde{R}_i - \tilde{C}_i)^{\text{def}}$	Ranking
CF1	0.1814	0.1814	5
CF2	0.1446	0.1446	14
CF3	0.1445	0.1445	15
CF4	0.1523	0.1523	13
CF5	0.1779	0.1779	6
CF6	0.1847	0.1847	3
CF7	0.1658	0.1658	10
CF8	0.2034	0.2034	2
CF9	0.3966	0.3966	1
CF10	0.1828	0.1828	4
CF11	0.1668	0.1668	8
CF12	0.1611	0.1611	11
CF13	0.1579	0.1579	12
CF14	0.1661	0.1661	9
CF15	0.1675	0.1675	7
FM1	0.1813	−0.1813	(8)
FM2	0.4412	−0.4412	(2)
FM3	0.5285	−0.5285	(1)
FM4	0.2033	−0.2033	(4)
FM5	0.2030	−0.2030	(5)
FM6	0.1935	−0.1935	(6)
FM7	0.1828	−0.1828	(7)
FM8	0.1668	−0.1668	(10)
FM9	0.1613	−0.1613	(11)
FM10	0.3238	−0.3238	(3)
FM11	0.1678	−0.1678	(9)

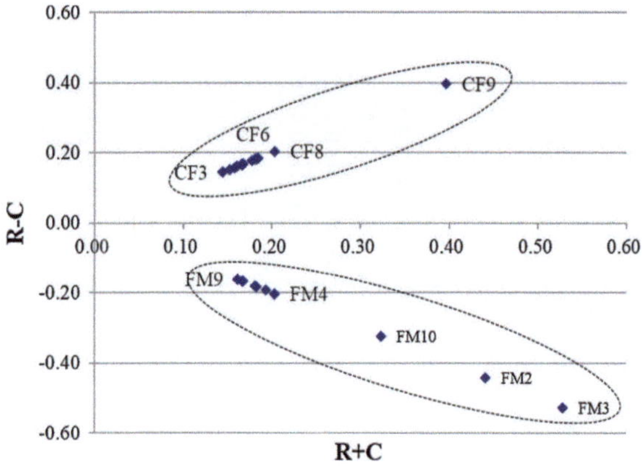

Fig. 9.7 Causal diagram of failure modes (Liu et al. 2015)

this will be followed by FM2, FM10, FM4, …, FM9. These results give a strong visualization to help decision makers carry out the risk evaluation, and hence, valuable cues can be obtained for identifying the most critical failure modes and assigning limited resources to them.

9.3.2 Comparisons and Discussion

To illustrate the effectiveness of the proposed FMEA method, the above case study is used to analyze some comparable methods, which include the conventional RPN method, the fuzzy OWA and DEMATEL method (Chang and Cheng 2011), and the OWGA and DEMATEL method (Chang 2009). Table 9.8 exhibits the ranking results of the fifteen failure causes as obtained using these approaches. By comparing the ranking results of the listed methods, the advantages that the proposed method has over other methods can be clearly seen.

First, from Table 9.8, we can find that except for CF1, CF3, and CF13, the risk priority rankings of other causes of failure obtained by the proposed method are different from those by the conventional RPN method. The main reasons for these differences can be explained by the shortcomings of the traditional FMEA introduced in Chap. 1. For example, according to the conventional RPN method, CF10 (RPN = 90) is assumed to be the most important and has a higher priority than CF9 (RPN = 70). However, the result of the proposed method shows that CF9 is the

Table 9.8 Ranking comparisons (Liu et al. 2015)

No.	Causes of failure	Proposed method	Traditional FMEA	Fuzzy OWA and DEMATEL ($\alpha = 0.7$)	OWGA and DEMATEL ($\alpha = 0.7$)
1	CF1	5	5	4	5
2	CF2	14	9	12	9
3	CF3	15	15	15	15
4	CF4	13	9	12	10
5	CF5	6	7	7	7
6	CF6	3	4	5	6
7	CF7	10	12	9	12
8	CF8	2	5	2	3
9	CF9	1	3	1	1
11	CF10	4	1	3	2
12	CF11	8	2	6	4
13	CF12	11	9	12	11
14	CF13	12	12	9	13
15	CF14	9	7	7	8
16	CF15	7	12	9	14

most important and has a higher priority compared with CF10, which tallied with the actual situation because the former has a higher severity and is therefore ranked higher than the latter. Similar situation can also be found between CF8 and CF11. In addition, in the case of more than one cause of failure having same RPN values, such as CF1 and CF8, CF5 and CF14, and CF2, CF4, and CF12, the proposed method can distinguish them from each other, thus providing more information than that of the traditional FMEA does. As expected, the proposed method considering the weights of risk factors achieves a more accurate risk priority ranking, discriminating among the results far more accurate than the conventional RPN method.

Second, there is much difference between the two sets of risk priority rankings produced by the proposed method and the fuzzy OWA and DEMATEL method. Also, the results obtained by the fuzzy OWA and DEMATEL method are not much discriminating. For example, CF2 cannot be distinguished from CF4 and CF12, which have exactly the same risk ranking. However, the result of the proposed method shows that among the three causes of failure, CF12 has the highest priority followed by CF4 and CF2. The main reasons for these differences are summarized as follows (Liu et al. 2015): (1) When each cause of failure is assigned to only one failure mode, the risk rankings obtained by the fuzzy OWA-based FMEA correspond with the ones obtained by the OWA operator. However, this mathematical formula for calculating the severity of influences among failure modes is questionable. (2) The fuzzy OWA and DEMATEL method simply used fuzzy values to define the risk factor ratings at the first phase of the method and then defuzzified them into crisp values to perform the DEMATEL procedure. This will lose some information when using the maximum membership degree to calculate the aggregated values by using the OWA weights. (3) The subjective weights of risk factors are not considered in the fuzzy OWA-based method, which may cause biased ranking results.

Third, the risk ranking order produced by the proposed method is different from the one by the OWGA and DEMATEL method. For example, only CF1, CF3, CF9, and CF12 are ranked in the same places. In addition, CF15 is ranked far behind CF14 although the former has a very high severity and should be ranked higher than the latter. The result of the proposed method shows that CF15 has a higher priority compared to CF14. These differences can be explained by the following reasons (Liu et al. 2015): (1) When each cause of failure is assigned to only one failure mode, the risk ranking orders obtained by the OWGA-based method and by the OWGA operator are the same. Similar to the fuzzy OWA operator, different sets of O, S, and D ratings can produce exactly the same aggregated value, but their hidden risk implications may be different. (2) The three risk factors are difficult to be precisely estimated by using the OWGA and DEMATEL method, which requires the risk factors for each failure mode to be precisely evaluated. This may not be realistic in real applications. (3) The OWGA-based FMEA ignores the subjective weights of risk factors and thus may result in biased conclusions.

The analysis of the results produced by the proposed method, the conventional RPN method, the fuzzy OWA and DEMATEL method, and the OWGA and DEMATEL method shows that a more accurate, reasonable risk assessment can be acquired by the application of the FWA and the fuzzy DEMATEL methods to FMEA. According to the domain experts, the proposed risk priority model is more suitable for the risk evaluation problem examined and can help decision makers find the most critical causes of failure effectively. Consequently, the reliability of the product can be assured by using the new risk assessment methodology. In comparison with the conventional RPN and its various improvements, the integration of FWA and fuzzy DEMATEL brings the following advantages: (1) The relative importance among risk factors is taken into consideration in the process of prioritization of failure modes. This makes the proposed fuzzy FMEA more practical and more flexible. (2) Risk factors and their relative importance weights are described as linguistic terms, which enable the experts to express their judgments more realistically and make the assessment easier to be carried out. (3) Both direct and indirect relationships between components of a system are considered in the priority setting process. Hence, the proposed fuzzy FMEA is a useful method for the prioritization of failure modes in complex systems with many subsystems or components. (4) The integrated method with FWA and fuzzy DEMATEL in fuzzy environment is new and has not appeared in the literature before. It provides a novel risk assessment methodology to help decision makers find the most critical failure causes and handle them by appropriate corrective actions in advance.

References

Chang KH (2009) Evaluate the orderings of risk for failure problems using a more general RPN methodology. Microelectron Reliab 49(12):1586–1596

Chang KH, Cheng CH (2011) Evaluating the risk of failure using the fuzzy OWA and DEMATEL method. J Intell Manuf 22(2):113–129

Chou C (2003) The canonical representation of multiplication operation on triangular fuzzy numbers. Comput Math Appl 45(10):1601–1610

Gabus A, Fontela E (1973) Perceptions of the World Problematique: Communication Procedure, Communicating With Those Bearing Collective Responsibility. DEMATEL Report No. 1 (in press). Battelle Geneva Research Centre, Geneva, Switzerland

Lin CJ, Wu WW (2004) A fuzzy extension of the DEMATEL method for group decision-making. Eur J Oper Res 156(1):445–455

Lin CJ, Wu WW (2008) A causal analytical method for group decision-making under fuzzy environment. Expert Syst Appl 34(1):205–213

Liou TS, Wang MJJ (1992) Fuzzy weighted average: an improved algorithm. Fuzzy Sets Syst 49(3):307–315

Liu HC, You JX, Lin QL, Li H (2015) Risk assessment in system FMEA combining fuzzy weighted average with fuzzy decision making trial and evaluation laboratory. Int J Comput Integr Manuf 28(7):701–714

Wu WW, Lee YT (2007) Developing global managers' competencies using the fuzzy DEMATEL method. Expert Syst Appl 32(2):499–507

Zadeh LA (1975) The concept of a linguistic variable and its application to approximate reasoning–I. Inf Sci 8(3):199–249

Chapter 10
FMEA Using Fuzzy Digraph and Matrix Approach

The digraph and matrix approach (Rao and Gandhi 2002; Baykasoglu 2014) is based on graph theory and matrix algebra and has some desirable properties, such as "ability to model criteria interactions" and "ability to generate hierarchical models," for solving complex decision-making problems. Considering the wide usage of fuzzy set theory and the advantages of digraph and matrix approach, Liu et al. (2014) proposed a new FMEA model, which uses fuzzy digraph and matrix approach for risk evaluation and prioritization of failure modes. The proposed FMEA can not only deal with the subjectivity and vagueness in both the risk factor weights determination and the failure modes evaluation, bust also visualize various risk factors and their interrelations using the graphical representation. Moreover, the assessments of failure modes and the relative importance of risk factors are used together to determine the risk priorities of the failure modes that have been identified in FMEA. The new model is able to make full use of the assessment information in risk analysis process, thus providing a more rational risk evaluation framework for FMEA.

10.1 Fuzzy Sets

Please refer to Sect. 11.1 for the basic definitions and operations related to fuzzy sets, such as triangular fuzzy number, linguistic variable, and the graded mean integration representation method, which will be applied in the proposed model.

H.-C. Liu, *FMEA Using Uncertainty Theories and MCDM Methods*,
DOI 10.1007/978-981-10-1466-6_10

10.2 Risk Factor Fuzzy Digraph and Matrix Representation

Risk factor is considered as an important factor which influences the determination of risk priority of the failure modes individuated in FMEA. *O*, *S*, and *D* are the three major risk factors utilized in FMEA. Conventionally, the three risk factors are evaluated in crisp values. Under many practical situations, however, it is hard, if not impossible, to obtain exact assessment values due to inherent vagueness and uncertainty in human judgments. As such, in this chapter, we choose linguistic terms for the assessment of risk factors. As the information about risk factors is fuzzy, its representation can also be implemented by a fuzzy digraph.

The risk factor fuzzy digraph models the risk factors and their interrelations. This digraph consists of a set of nodes $V = (v_j)$, with $j = 1, 2, \ldots, n$ and a set of direct edges $E = (e_{jk})$. A node v_j represents the *j*th risk factor, RF_j, and edges represent the relative importance between the risk factors which can be expressed by fuzzy numbers. The number of nodes *n* is considered to be equal to the number of risk factors defined in FMEA. If the risk factor RF_j is having relative importance over another risk factor RF_k in the FMEA process, then a directed edge or arrow is drawn from node *j* to node *k*, i.e., e_{jk}. Similarly, if RF_k is having relative importance over RF_j, then a directed edge or arrow is drawn from node *k* to node *j*, i.e., e_{kj}. The fuzzy digraph, based on risk factors and their interrelations, is shown in Fig. 10.1.

The risk factor fuzzy digraph gives a graphical representation of risk factors and their relative importance for quick visual appraisal. However, as the number of risk factors and their interrelations increases, the digraph becomes complex and its

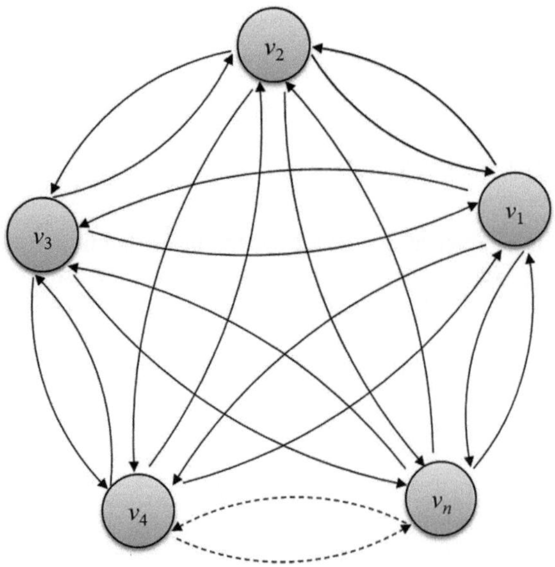

Fig. 10.1 Risk factor fuzzy digraph (Liu et al. 2014)

visual analysis is expected to be difficult and complex. To overcome this constraint, the digraph can be represented in a matrix form.

Matrix representation of the risk factor fuzzy digraph offers one-to-one representation. A matrix called the risk factor fuzzy matrix is defined. Generally, if there are n risk factors RF_j $(j = 1, 2, \ldots, n)$ and relative importance exists between all of the risk factors, then the risk factor fuzzy matrix $\tilde{A}$ for the risk factor fuzzy digraph in Fig. 10.1 can be represented as

$$\tilde{A} = \begin{bmatrix} \tilde{A}_1 & \tilde{a}_{12} & \tilde{a}_{13} & \ldots & \tilde{a}_{1n} \\ \tilde{a}_{21} & \tilde{A}_2 & \tilde{r}_{23} & \ldots & \tilde{a}_{2n} \\ \tilde{a}_{31} & \tilde{a}_{32} & \tilde{A}_3 & \ldots & \tilde{a}_{3n} \\ \vdots & \vdots & \vdots & \ldots & \vdots \\ \tilde{a}_{n1} & \tilde{a}_{n2} & \tilde{a}_{n3} & \ldots & \tilde{A}_n \end{bmatrix}, \tag{10.1}$$

where $\tilde{A}_j$ is the fuzzy value of RF_j represented by node v_j, and $\tilde{a}_{jk}$ is the fuzzy relative importance of RF_j over RF_k represented by the edge e_{jk}. The permanent of this fuzzy matrix $\tilde{A}$, $\mathrm{per}(\tilde{A})$, is defined as the risk priority function. The permanent is a standard matrix function and is used in combinatorial mathematics (Baykasoglu 2014). Application of permanent concept in FMEA will help in representing risk factors of failure modes from combinatorial consideration, thus leading to a better evaluation of failure modes. Moreover, no information will be lost in the calculation process because no negative sign will appear in the permanent function of a matrix (Rao and Padmanabhan 2006; Liu et al. 2014). The risk priority function characterizes the considered risk analysis problem as it contains all possible structural components of risk factors and their relative importance.

10.3 The Proposed Model for FMEA

The proposed FMEA model is based on fuzzy digraph and matrix approach considering the interrelations of risk factors for the given criticality analysis problem. Different from the conventional RPN method, the new FMEA model treats the risk factors and their relative weights as fuzzy variables and assesses them using fuzzy linguistic terms and fuzzy grades. For example, the linguistic values from the linguistic term set (*Very Low*, *Low*, *Medium*, *High*, and *Very High*) are applied in this chapter for the description of the importance weights of risk factors (Table 10.1). Similarly, the linguistic variables from the linguistic term set (*Very Low*, *Low*, *Medium Low*, *Medium*, *Medium High*, *High*, and *Very High*) are used for the description of the fuzzy ratings of failure modes with respect to each risk factor (Table 10.2). The membership functions of the two sets of linguistic values are shown in Figs. 10.2 and 10.3, respectively.

Table 10.1 Linguistic terms for rating risk factor weights

Linguistic terms	Fuzzy numbers
Very low (VL)	(0, 0.1, 0.3)
Low (L)	(0.1, 0.3, 0.5)
Medium (M)	(0.3, 0.5, 0.7)
High (H)	(0.5, 0.7, 0.9)
Very high (VH)	(0.7, 0.9,1.0)

Table 10.2 Linguistic terms for rating failure modes

Linguistic terms	Fuzzy numbers
Very low (VL)	(0, 0, 1)
Low (L)	(0, 1, 2)
Medium low (ML)	(1, 3, 5)
Medium (M)	(3, 5, 7)
Medium high (MH)	(5, 7, 9)
High (H)	(8, 9, 10)
Very high (VH)	(9, 10, 10)

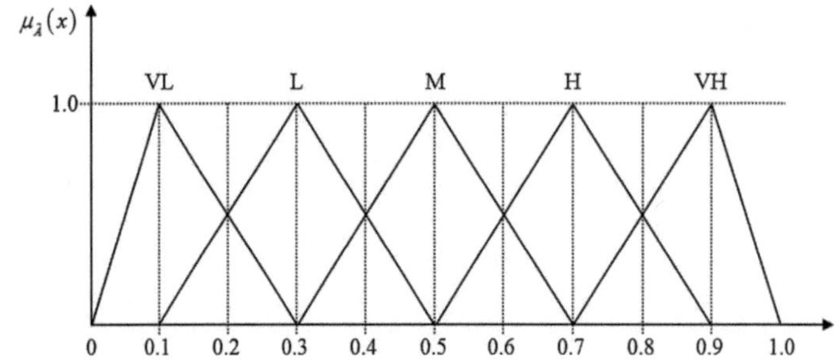

Fig. 10.2 Membership functions for rating risk factor weights

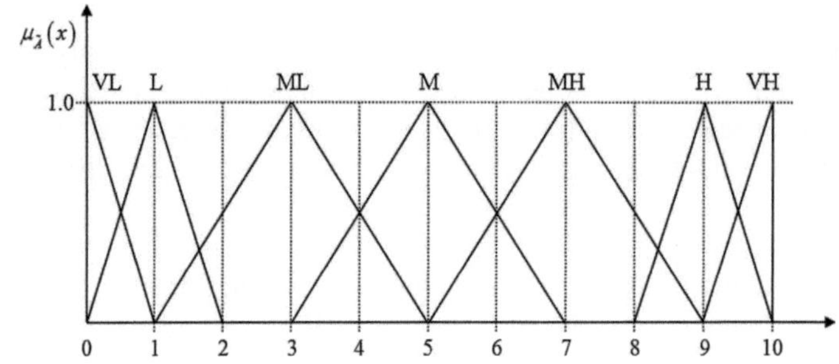

Fig. 10.3 Membership functions for rating failure modes

After aggregating FMEA team members' individual evaluations into group assessments, the fuzzy digraph and matrix approach is applied to determine the interrelation matrix of risk factors and construct the fuzzy risk matrix for each failure mode. Finally, the risk priority index (RPI) derived from the risk priority function is utilized to determine the risk priority of all the recognized failure modes. The flow diagram in Fig. 10.4 shows the overall procedure for ranking the failure modes in FMEA using the proposed model.

Suppose that there are l cross-functional team members TM_k $(k = 1, 2, \ldots, l)$ in a FMEA team responsible for the assessment of m failure modes FM_i $(i = 1, 2, \ldots, m)$ with respect to n risk factors RF_j $(j = 1, 2, \ldots, n)$. Each team member

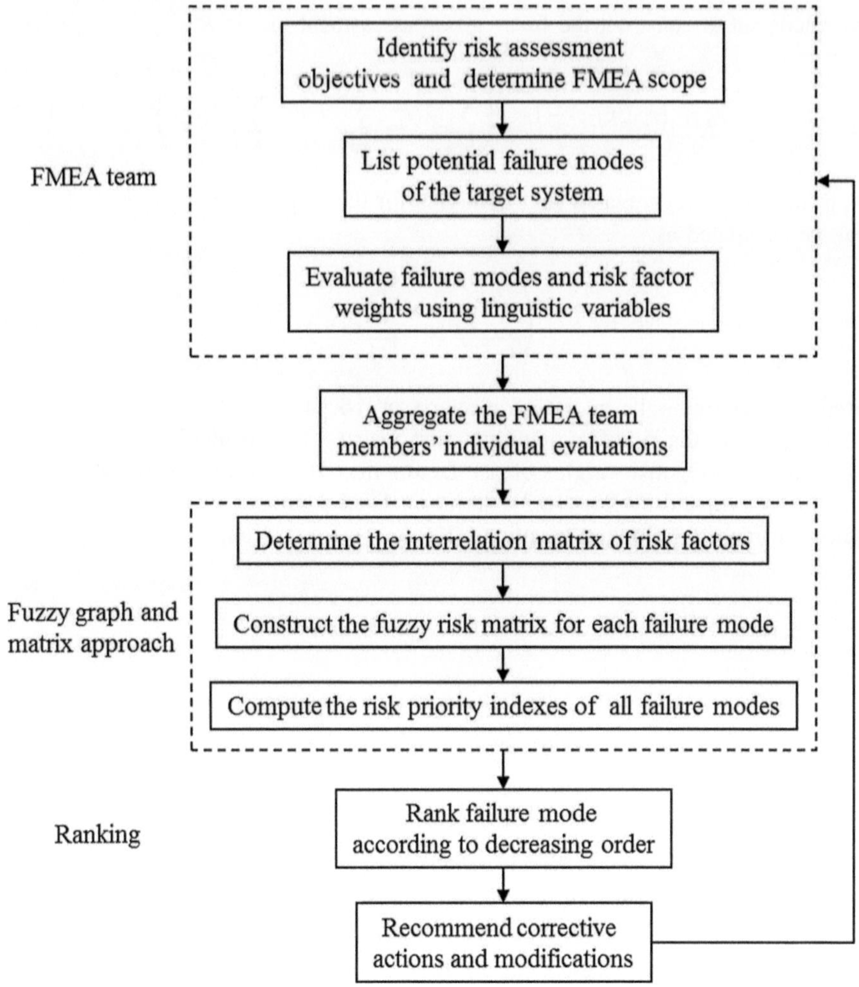

Fig. 10.4 Flowchart of the proposed FMEA model (Liu et al. 2014)

TM_k is given a weight $\lambda_k > 0$ $(k = 1, 2, \ldots, l)$ satisfying $\sum_{k=1}^{l} \lambda_k = 1$ to reflect his/her relative importance in the FMEA process. Let $\tilde{R}_k = \left(\tilde{r}_{ij}^k\right)_{m \times n}$ be the fuzzy assessment matrix of the kth team member, where $\tilde{r}_{ij}^k = \left(r_{ij1}^k, r_{ij2}^k, r_{ij3}^k\right)$ is the fuzzy rating provided by TM_k on the assessment of FM_i with respect to RF_j. Let $\tilde{w}_j^k = \left(w_{j1}^k, w_{j2}^k, w_{j3}^k\right)$ is the fuzzy weight of the risk factor RF_j given by TM_k to reflect its relative importance in determining the risk priority ranking of failure modes. Based upon these assumptions and notations, the m failure modes can be prioritized through the following steps (Liu et al. 2014):

Step 1. Aggregate the FMEA team members' individual evaluations

The aggregated fuzzy ratings of failure modes with respect to each risk factor can be calculated to construct the fuzzy group assessment matrix $\tilde{R} = \left(\tilde{r}_{ij}\right)_{m \times n}$, where

$$\tilde{r}_{ij} = \left(r_{ij1}, r_{ij2}, r_{ij3}\right) = \left(\sum_{k=1}^{l} \lambda_k r_{ij1}^k, \sum_{k=1}^{l} \lambda_k r_{ij2}^k, \sum_{k=1}^{l} \lambda_k r_{ij3}^k\right). \tag{10.2}$$

Similarly, the aggregated fuzzy weights for the n risk factors $\tilde{w}_j (j = 1, 2, \ldots n)$ can be calculated as:

$$\tilde{w}_j = \left(w_{j1}, w_{j2}, w_{j3}\right) = \left(\sum_{k=1}^{l} \lambda_k w_{j1}^k, \sum_{k=1}^{l} \lambda_k w_{j2}^k, \sum_{k=1}^{l} \lambda_k w_{j3}^k\right). \tag{10.3}$$

Step 2. Determine the interrelation matrix of risk factors

To generate the risk factor interrelation matrix, it is required to calculate the normalized aggregated weights of risk factors first. Based on the aggregated fuzzy weights determined by Eq. (10.3), the normalized aggregated weights can be calculated using the following equations (Koulouriotis and Ketipi 2011):

$$\tilde{v}_j = \frac{\tilde{w}_j}{\sum_{j=1}^{n} w_{j3}}, \tag{10.4}$$

$$\tilde{w'}_j = \left(w'_{j1}, w'_{j2}, w'_{j3}\right) = \frac{\tilde{v}_j}{\sum_{j=1}^{n} \bar{v}_j}. \tag{10.5}$$

where $\bar{v}_j$ is the defuzzified value of $\tilde{v}_j$ and $\sum_{j=1}^{n} \bar{w}'_j = 1$.

The interrelation matrix of risk factors results from the interrelations represented by the risk factor fuzzy digraph. For n risk factors, their relative importance relations can be expressed in the form of interrelation matrix $\widetilde{\mathrm{IM}}$ as

$$\widetilde{\mathrm{IM}} = \begin{bmatrix} \tilde{I}_1 & \tilde{a}_{12} & \tilde{a}_{13} & \cdots & \tilde{a}_{1n} \\ \tilde{a}_{21} & \tilde{I}_2 & \tilde{r}_{23} & \cdots & \tilde{a}_{2n} \\ \tilde{a}_{31} & \tilde{a}_{32} & \tilde{I}_3 & \cdots & \tilde{a}_{3n} \\ \vdots & \vdots & \vdots & \cdots & \vdots \\ \tilde{a}_{n1} & \tilde{a}_{n2} & \tilde{a}_{n3} & \cdots & \tilde{I}_n \end{bmatrix}, \tag{10.6}$$

where $\tilde{I}_j = (1,1,1)$, $j = 1,2,\ldots n$. The importance of risk factor RF_p relatively to risk factor RF_q, $\tilde{a}_{pq}$, and the relative importance $\tilde{a}_{qp}$ can be computed by the following expressions (Koulouriotis and Ketipi 2011):

$$\tilde{u}_p = \frac{\tilde{w}'_p}{w_{p3} + w_{q3}}, \quad \tilde{u}_q = \frac{\tilde{w}'_q}{w_{p3} + w_{q3}}, \tag{10.7}$$

$$\tilde{a}_{pq} = \frac{\tilde{u}_p}{\bar{u}_p + \bar{u}_q}, \quad \tilde{a}_{qp} = \frac{\tilde{u}_q}{\bar{u}_p + \bar{u}_q}. \tag{10.8}$$

where $\tilde{w}'_p = \left(w'_{p1}, w'_{p2}, w'_{p3}\right)$, and $\tilde{w}'_q = \left(w'_{q1}, w'_{q2}, w'_{q3}\right)$ are the normalized aggregated weights of risk factors RF_p and RF_q, respectively, $\bar{u}_p$ and $\bar{u}_q$ are the defuzzified values of $\tilde{u}_p$ and $\tilde{u}_q$ which can be calculated by using Eq. (10.7), and $\bar{a}_{pq} + \bar{a}_{qp} = 1$.

Step 3. Construct the fuzzy risk matrix for each failure mode

To develop the fuzzy risk matrix for each failure mode, the main diagonal of the interrelation matrix is substituted by the normalized aggregated fuzzy ratings of the corresponding risk factors for the particular failure mode. Thus, while we are referred to the failure mode FM_i and the n risk factors, the element $\tilde{I}_j$ can be replaced with the normalized aggregated rating $\tilde{r}'_{ij}$ as seen below:

$$\widetilde{\mathrm{RM}}_i = \begin{bmatrix} \tilde{r}'_{i1} & \tilde{a}_{12} & \tilde{a}_{13} & \cdots & \tilde{a}_{1n} \\ \tilde{a}_{21} & \tilde{r}'_{i2} & \tilde{r}_{23} & \cdots & \tilde{a}_{2n} \\ \tilde{a}_{31} & \tilde{a}_{32} & \tilde{r}'_{i3} & \cdots & \tilde{a}_{3n} \\ \vdots & \vdots & \vdots & \cdots & \vdots \\ \tilde{a}_{n1} & \tilde{a}_{n2} & \tilde{a}_{n3} & \cdots & \tilde{r}'_{in} \end{bmatrix}. \tag{10.9}$$

Here, the linear scale transformation method of Chen (2000) is used to normalize the aggregated fuzzy ratings, which is shown in Eq. (10.10).

$$\tilde{r}'_{ij} = = \left(\frac{r_{ij1}}{\max_i r_{ij3}}, \frac{r_{ij2}}{\max_i r_{ij3}}, \frac{r_{ij3}}{\max_i r_{ij3}} \right). \tag{10.10}$$

The normalization method mentioned above is to preserve the property that the range of the normalized triangular fuzzy numbers belongs to [0, 1].

Step 4. Compute the risk priority index (RPI)

The permanent of the fuzzy risk matrix, $\mathrm{per}\left(\widetilde{\mathrm{RM}}_i\right)$, is defuzzified by using the graded mean integration representation method to obtain the risk priority index of each failure mode RPI_i in order to evaluate and rank the failure modes.

Step 5. Determine the ranking order of all failure modes

The higher the risk priority index RPI_i, the greater risk of the failure mode, and the higher the risk priority. Therefore, the risk priority of all the failure modes can be determined in terms of their risk priority indexes.

Step 6. Analyze the results and develop recommendations to enhance the system performance

Having obtained the risk ranking of failure modes, risk managers can select the failure modes with the greatest risk among a set of identified failure modes to arrange improvement resources and take remedial actions.

10.4 An Illustrative Example

10.4.1 Implementation

In what follows, a case study of steam valve system in a power generation plant (Liu et al. 2014; Song et al. 2014) is provided to demonstrate the proposed fuzzy FMEA. Steam valve system is a key part of the steam turbine operation, which can affect the reliability of the whole power plant if it fails. The steam valve system is set between moisture separator reheater (MSR) and low-pressure cylinder to control transporting medium with high temperature and pressure, and operates in complex working conditions. The steam valve must be fully opened and closed timely, and it must work stably in any position from full-open to full-closed. Therefore, to ensure the safety and reliability of the turbine system, steam valve is required to be quickly closed in any abnormal operating circumstances to cut off the steam entering into the low-pressure cylinder. First, an FMEA team of four experts, $\mathrm{TM}_k (k = 1, 2, 3, 4)$, has been formed to conduct the risk evaluation and to identify the most serious failure modes. Eight major potential failure modes were explored and listed by the FMEA team through brainstorming. These failure modes, the causes leading to them, their possible effects, and detection measures are presented in Table 10.3. The four team members are assigned the following relative weights: 0.15, 0.30, 0.35, and 0.20 in performing the FMEA analysis because of their different domain knowledge and experience.

The risk factors O, S, and D are chosen for the prioritization of the identified failure modes. The fuzzy assessment matrix from experts will be analyzed with the aid of the proposed FMEA model to rank the risk of the identified failure modes. The FMEA team members use the linguistic terms shown in Table 10.1 to evaluate the relative importance of the risk factors, and employ the linguistic terms shown in Table 10.2 to evaluate the risk factors for each failure mode. The results are provided in Tables 10.4 and 10.5, respectively.

Table 10.3 FMEA of the steam valve system (Song et al. 2014; Liu et al. 2014)

No.	Failure modes	Causes of failure	Effects of failure	Detection measures
FM1	Long closing time of valve	Unreasonable spring selection	Over speeding of steam turbine rotor and parts breakdown	Valve closing test
FM2	Not being tightly closed	Small bushing clearance, shaft bending	Blade corrosion of steam turbine	Valve leak test
FM3	Steam leak around valve shaft	Compaction force of sealing filler is not enough	Waste of chemical water and thermal loss	Inspection after packing removal
FM4	Valve fluctuations	Hydraulic cylinder leaks	Valve cannot open and close normally, and unsafe operation	Visual inspection of cylinder pressure gauge
FM5	Valve jam in operation	Large deformation of valve shaft or body due to process and material defects	Valve cannot open and close affecting normal operation of turbine	Valve operation test
FM6	Fracture of valve shaft	Fatigue fracture under alternating stress	Tripping of turbine unit	Metallographic tests on the fracture gap
FM7	Malfunction of valve shaft support bearing	Low strength of bearing material and long-term wear and tear	Abnormal operation of valve system	Disassemble inspection
FM8	Excessive noise of valve system	System vibration due to unreasonable components clearance selection	Make the user feel uncomfortable and reduce service life of components	Change operating condition, frequency measurement of valve system

Table 10.4 Linguistic evaluations of risk factor weights (Liu et al. 2014)

Risk factors	Team members			
	TM1	TM2	TM3	TM4
O	L	M	M	H
S	M	H	H	VH
D	M	H	M	H

After transformed into respective triangular fuzzy numbers, the FMEA team members' individual evaluations are aggregated using Eqs. (10.2) and (10.3) to construct the fuzzy group assessment matrix $\tilde{R} = \left(\tilde{r}_{ij}\right)_{8\times 3}$ and to obtain the aggregated fuzzy weights for the three risk factors $\tilde{w}_j (j = 1, 2, 3)$. The results are shown in Table 10.6.

Table 10.5 Linguistic evaluations of failure modes (Liu et al. 2014)

Risk factors		*O*				*S*				*D*			
Team members		TM1	TM2	TM3	TM4	TM1	TM2	TM3	TM4	TM1	TM2	TM3	TM4
Failure modes	FM1	ML	ML	M	L	VH	VH	VH	H	H	H	MH	M
	FM2	ML	ML	M	ML	L	VL	L	ML	ML	M	M	ML
	FM3	MH	MH	M	M	ML	ML	M	ML	ML	ML	L	M
	FM4	ML	L	ML	ML	MH	MH	MH	M	M	L	ML	ML
	FM5	ML	ML	M	M	VH	VH	MH	H	ML	ML	L	ML
	FM6	L	ML	L	ML	VH	VH	VH	VH	L	ML	ML	M
	FM7	M	M	MH	M	MH	MH	H	H	ML	L	ML	ML
	FM8	M	MH	M	MH	M	M	MH	MH	ML	M	ML	ML

Table 10.6 Fuzzy group assessment matrix and aggregated fuzzy weights of risk factors (Liu et al. 2014)

Failure modes	*O*	*S*	*D*
FM1	(1.5, 3.3, 5.1)	(8.8, 9.8, 10)	(5.95, 7.5, 9.05)
FM2	(1.7, 3.7, 5.7)	(0.2, 1.1, 2.3)	(2.3, 4.3, 6.3)
FM3	(3.9, 5.9, 7.9)	(1.7, 3.7, 5.7)	(1.05, 2.7, 4.35)
FM4	(0.7, 2.4, 4.1)	(4.6, 6.6, 8.6)	(1, 2.7, 4.4)
FM5	(2.1, 4.1, 6.1)	(7.4, 8.75, 9.65)	(0.65, 2.3, 3.95)
FM6	(0.5, 2, 3.5)	(9, 10, 10)	(1.25, 3.1, 4.95)
FM7	(3.7, 5.7, 7.7)	(6.65, 8.1, 9.55)	(0.7, 2.4, 4.1)
FM8	(4, 6, 8)	(4.1, 6.1, 8.1)	(1.6, 3.6, 5.6)
Weights	(0.31, 0.51, 0.71)	(0.51, 0.71, 0.89)	(0.4, 0.6, 0.8)

The following step is the calculation of the interrelation matrix of risk factors. Considering the aggregated fuzzy risk factor weights illustrated in Table 10.6 and using Eqs. (10.4)–(10.6), the formation of interrelation matrix $\widetilde{\text{IM}}$ is easy to be accomplished as expressed in Table 10.7. Each element of this matrix represents the relative importance between two risk factors.

Subsequently, utilizing Eq. (10.10), the normalized aggregated ratings of risk factors are calculated as shown in Table 10.8. As can be noticed from Table 10.7, the interrelation matrix's main diagonal consists of ones. Replacing the ones of the diagonal with the normalized aggregated ratings, a fuzzy risk matrix $\widetilde{\text{RM}}_i$ is created for each failure mode as shown at Eq. (10.9).

Table 10.7 The interrelation matrix of risk factors (Liu et al. 2014)

Risk factors	*O*	*S*	*D*
O	(1, 1, 1)	(0.255, 0.419, 0.584)	(0.279, 0.459, 0.640)
S	(0.419, 0.584, 0.732)	(1, 1, 1)	(0.390, 0.543, 0.681)
D	(0.360, 0.541, 0.721)	(0.306, 0.459, 0.612)	(1, 1, 1)

Table 10.8 The normalized aggregated ratings of risk factors (Liu et al. 2014)

Failure modes	*O*	*S*	*D*
FM1	(0.188, 0.413, 0.638)	(0.880, 0.980, 1.000)	(0.657, 0.829, 1.000)
FM2	(0.213, 0.463, 0.713)	(0.020, 0.110, 0.230)	(0.254, 0.475, 0.696)
FM3	(0.488, 0.738, 0.988)	(0.170, 0.370, 0.570)	(0.116, 0.298, 0.481)
FM4	(0.088, 0.300, 0.513)	(0.460, 0.660, 0.860)	(0.110, 0.298, 0.486)
FM5	(0.263, 0.513, 0.763)	(0.740, 0.875, 0.965)	(0.072, 0.254, 0.436)
FM6	(0.063, 0.250, 0.438)	(0.900, 1.000, 1.000)	(0.138, 0.343, 0.547)
FM7	(0.463, 0.713, 0.963)	(0.665, 0.810, 0.955)	(0.077, 0.265, 0.453)
FM8	(0.500, 0.750, 1.000)	(0.410, 0.610, 0.810)	(0.177, 0.398, 0.619)

Table 10.9 Fuzzy permanents and risk priority indexes (Liu et al. 2014)

Failure modes	Fuzzy permanent	RPI	Ranking
FM1	(0.361, 1.130, 2.364)	1.2078	1
FM2	(0.127, 0.529, 1.387)	0.6053	8
FM3	(0.169, 0.677, 1.723)	0.7664	6
FM4	(0.145, 0.617, 1.605)	0.7030	7
FM5	(0.199, 0.768, 1.843)	0.8521	4
FM6	(0.192, 0.726, 1.689)	0.7978	5
FM7	(0.226, 0.843, 2.024)	0.9371	3
FM8	(0.228, 0.864, 2.129)	0.9689	2

Next, the permanents of the fuzzy risk matrixes $\widetilde{\mathrm{RM}}_i$ $(i = 1, 2, \ldots, 8)$ are computed for the eight failure modes. Since the failure mode ratings and the risk factor weights are all fuzzy numbers, the resulted permanents for the failure modes are also fuzzy numbers as presented in Table 10.9. Finally, the risk priority index of each failure mode RPI_i is obtained by defuzzifing the fuzzy permanents into crisp values with the graded mean integration representation method.

As shown in Table 10.9, FM1 is apparently the failure mode with the maximum overall risk and should be given the top priority for correction, followed by FM8, FM7, FM5, FM6, FM3, FM4, and FM2.

10.4.2 Comparisons and Discussion

To verify the effectiveness of the proposed model, the conventional RPN method and two other FMEA methodologies have been implemented for the given case study. These methodologies are the rough TOPSIS (Song et al. 2014) and the fuzzy VIKOR (Liu et al. 2012). The risk ranking of the eight failure modes which has resulted for each method is presented in Table 10.10.

Table 10.10 Ranking comparisons (Liu et al. 2014)

Failure modes	The proposed model	The conventional RPN method					The rough TOPSIS		The fuzzy VIKOR	
		O	*S*	*D*	RPN	Ranking	CC_i	Ranking	Q_i	Ranking
FM1	1	4	9	7	252	1	0.367	1	0.960	1
FM2	8	4	3	4	48	8	0.737	8	0.000	8
FM3	6	6	4	4	96	6	0.600	7	0.506	6
FM4	7	4	6	4	96	6	0.587	6	0.209	7
FM5	4	4	9	4	144	3	0.474	3	0.516	5
FM6	5	4	10	3	120	4	0.493	4	0.694	2
FM7	3	7	8	2	112	5	0.473	2	0.600	4
FM8	2	6	6	5	180	2	0.495	5	0.642	3

From Table 10.10, it can be observed that the ranking orders of the first two failure modes are exactly the same for all the four methods; i.e., FM1 is with the maximum overall risk, whereas FM2 is with the least overall risk and should be given the lowest risk priority. However, most of the failure modes have different ranks in the four different FMEA approaches. This shows that the four FMEA methods (the proposed model, the conventional RPN method, the rough TOPSIS-based FMEA, and the fuzzy VIKOR-based FMEA) have different mechanisms in determining the risk priority ranking of failure modes. Consider FM3 and FM4, where the RPN is 96. Although the RPNs for the two failures are the same, their risk levels are different. The result of the proposed model shows that FM3 has a higher risk compared to FM4, which is consistent with the result obtained by the fuzzy VIKOR. This can be understood from the fact that in the traditional FMEA, the risk factors O, S, and D are evaluated in a precise way and simply multiplied to produce the RPN without considering their weights and the uncertainty of experts' subjective evaluations. In addition, the result of the proposed model shows that FM5 has a higher risk in comparison with FM6. Both the conventional RPN and the rough TOPSIS methods also give a higher rank to FM5. However, the fuzzy VIKOR-based FMEA puts FM6 a very high priority. As for FM7 and FM8, the former failure mode is the second most important failure and has a higher risk than the latter one by the proposed fuzzy FMEA. This can be collated by the conventional RPN and the fuzzy VIKOR methods, which assign higher importance to FM7. However, FM8 ranks the second and FM7 is ranked far behind FM8 when the rough TOPSIS-based FMEA is applied. The main reason for these inconsistencies is that the interrelations between risk factors are not considered in the fuzzy VIKOR and the rough TOPSIS, which may cause biased ranking results.

From the above analysis, it can be concluded that the risk priority ranking of failure modes given by the proposed model is more accurate and reliable. Compared with the convention RPN method and its various improvements, the proposed FMEA model using fuzzy digraph and matrix approach has the following advantages: (1) It allows experts to evaluate the risk factors and their relative weights using linguistic terms rather than in a precise way as the traditional FMEA. (2) The assessments of risk factors and their relative importance are used together to rank the failure modes and hence, it provides a better evaluation of the failure modes. The proposed fuzzy FMEA characterizes the considered risk analysis problem as it contains all possible structural components of risk factors and their relative importance. (3) The proposed model not only provides an analysis of failure modes, but also enables the visualization of the risk factors and their interrelations through the graphical representation. (4) A small variation in each of the risk factors can lead to a significant difference in the risk priority index and hence, the proposed FMEA algorithm is able to fully prioritize failure modes with clear-cut difference in their risk priority indexes. (5) The model proposed in this chapter is simple and effective for FMEA and in particular, the defined RPIs offer a new way for prioritizing failure modes in FMEA.

References

Baykasoglu A (2014) A review and analysis of "graph theoretical-matrix permanent" approach to decision making with example applications. Artif Intell Rev 42(4):573–605

Chen CT (2000) Extensions of the TOPSIS for group decision-making under fuzzy environment. Fuzzy Sets Syst 114(1):1–9

Koulouriotis DE, Ketipi MK (2011) A fuzzy digraph method for robot evaluation and selection. Expert Syst Appl 38(9):11901–11910

Liu HC, Liu L, Liu N, Mao LX (2012) Risk evaluation in failure mode and effects analysis with extended VIKOR method under fuzzy environment. Expert Syst Appl 39(17):12926–12934

Liu HC, Chen YZ, You JX, Li H (2014) Risk evaluation in failure mode and effects analysis using fuzzy digraph and matrix approach. J Intell Manuf. doi:10.1007/s10845-014-0915-6

Rao RV, Gandhi OP (2002) Failure cause analysis of machine tools using digraph and matrix methods. Int J Mach Tools Manuf 42(4):521–528

Rao RV, Padmanabhan KK (2006) Selection, identification and comparison of industrial robots using digraph and matrix methods. Robot Comput Integr Manuf 22(4):373–383

Song W, Ming X, Wu Z, Zhu B (2014) A rough TOPSIS approach for failure mode and effects analysis in uncertain environments. Qual Reliab Eng Int 30(4):473–486

Chapter 11
FMEA Using Fuzzy MULTIMOORA Method

The MULTIMOORA method (Brauers and Zavadskas 2010) is a recently introduced MCDM method based on the multi-objective optimization by ratio analysis (MOORA) (Brauers and Zavadskas 2006). Due to its characteristics and capabilities, the use of MULTIMOORA method has been increasing in the literature. In Liu et al. (2014), the authors proposed a new risk priority model by applying fuzzy set theory and MULTIMOORA method for failure modes assessment and ranking in FMEA. The risk factors and their relative weights are treated as fuzzy variables and evaluated by using fuzzy linguistic terms and fuzzy ratings. An extended MULTIMOORA method is used to determine the risk ranking of the failure modes that have been identified. The new risk priority model can be a useful tool for determining the ranking orders of the identified failure modes in FMEA and taking preventive actions for safety and reliability improvement.

11.1 Fuzzy Set Theory and MULTIMOORA Method

11.1.1 Fuzzy Set Theory

The basic definitions of fuzzy sets can be found in Sect. 7.1.1, and, in the following, we only introduce the distance between trapezoidal fuzzy numbers which will be utilized in the FMEA model proposed in this chapter.

Definition 11.1 Let $\tilde{A} = (a_1, a_2, a_3, a_4)$ and $\tilde{B} = (b_1, b_2, b_3, b_4)$ be two trapezoidal fuzzy numbers, then the distance between them can be calculated by using the vertex method as (Wan and Li 2013; Liu et al. 2014):

$$d(\tilde{A}, \tilde{B}) = \sqrt{\frac{1}{6}\left[(a_1 - b_1)^2 + 2(a_2 - b_2)^2 + 2(a_3 - b_3)^2 + (a_4 - b_4)^2\right]}. \quad (11.1)$$

H.-C. Liu, *FMEA Using Uncertainty Theories and MCDM Methods*,
DOI 10.1007/978-981-10-1466-6_11

11.1.2 The MULTIMOORA Method

The MULTIMOORA method introduced by Brauers and Zavadskas (2006) begins with a decision matrix X where its elements x_{ij} denote the values of the ith alternative on the jth criterion (objective), $i = 1, 2, \ldots, m$ and $j = 1, 2, \ldots, n$. It consists of three parts: the ratio system, the reference point approach, and the full multiplicative form.

The Ratio System. Ratio system employs the vector data normalization by comparing an alternative of a criterion to all values of criteria.

$$x_{ij}^* = \frac{x_{ij}}{\sqrt{\sum_{i=1}^{m} x_{ij}^2}}, \tag{11.2}$$

where x_{ij}^* represents the normalized value of the ith alternative on the jth criterion. These normalized values are added (if desirable value of criterion is maximum) or subtracted (if desirable value is minimum). Thus, the summarizing index of each alternative is derived by

$$y_i^* = \sum_{j=1}^{g} x_{ij}^* - \sum_{j=g+1}^{n} x_{ij}^*, \tag{11.3}$$

where $g = 1, 2, \ldots, n$ denotes number of criteria to be maximized and y_i^* is the overall assessment of the ith alternative with respect to all criteria. Then, the rank of alternatives is given according to every summarizing index: The higher the index, the higher the rank.

The Reference Point Approach. Reference point approach is based on the ratio system. The maximal objective reference point (MORP) is found according to the ratios computed by Eq. (11.2). The jth coordinate of the reference point can be described as $r_j = \max_i x_{ij}^*$ in the case of maximization. Every coordinate of this vector represents maximum or minimum of certain criterion. Then, the ranking of alternatives is given according to the deviation from the reference point and the min–max Metric of Tchebycheff:

$$\min_i \left\{ \max_j \left| r_j - x_{ij}^* \right| \right\}. \tag{11.4}$$

The Full Multiplicative Form. The full multiplicative form method embodies maximization as well as minimization of purely multiplicative utility function. The overall utility of the ith alternative can be expressed as dimensionless number by using the following relation:

$$U_i = \frac{A_i}{B_i}, \tag{11.5}$$

where $A_i = \prod_{j=1}^{g} x_{ij}$ is the product of criteria of the ith alternative to be maximized and $B_i = \prod_{j=g+1}^{n} x_{ij}$ denotes the product of criteria of the ith alternative to be minimized.

The Dominance Theory. Brauers and Zavadskas (2011) developed the theory of dominance to summarize the three rank lists provided by different parts of MULTIMOORA into a single one. For detailed information regarding the dominance theory, readers can refer to Brauers and Zavadskas (2011, 2012).

11.2 The Proposed Model for FMEA

In this section, a systematic approach to extend the MULTIMOORA method is proposed to assess the risk of potential failure modes in the fuzzy environment. The flow diagram in Fig. 11.1 shows the proposed approach to rank the identified failure modes in FMEA process.

Suppose there are l cross-functional members $\text{TM}_k(k = 1, 2, \ldots, l)$ in a FMEA team responsible for the assessment of m failure modes $\text{FM}_i(i = 1, 2, \ldots, m)$ with respect to n risk factors $\text{RF}_j(j = 1, 2, \ldots, n)$. Each team member TM_k is given a weight $\lambda_k > 0(k = 1, 2, \ldots, l)$ satisfying $\sum_{k=1}^{l} \lambda_k = 1$ to reflect his/her relative importance in the risk analysis process. Let $\tilde{R}_k = \left(\tilde{r}_{ij}^k\right)_{m \times n}$ be the fuzzy assessment matrix of the kth team member, where $\tilde{r}_{ij}^k = \left(r_{ij1}^k, r_{ij2}^k, r_{ij3}^k, r_{ij4}^k\right)$ is the fuzzy rating provided by TM_k on the assessment of FM_i with respect to RF_j. Let $\tilde{w}_j^k = \left(w_{j1}^k, w_{j2}^k, w_{j3}^k, w_{j4}^k\right)$ is the fuzzy weight of the risk factor RF_j given by TM_k to reflect its relative importance in the risk ranking of the identified failure modes. Based on the above, the procedure of the proposed FMEA model can be summarized as the following steps (Liu et al. 2014):

Step 1. Identify the objectives of risk assessment and define FMEA scope

The first step is defining the objectives of risk assessment. Giving clear and careful thought to this step is very critical to the following risk evaluation and ranking process. Then, the scope of failure analysis problem should be well defined, usually by the leader of the function responsible for the FMEA. A specific and clear definition of the process or product to be analyzed will help prevent the team from focusing on the wrong aspects of the product or process during the FMEA.

Step 2. Assemble a FMEA team and list all potential failure modes

As mentioned previously, in FMEA process several decision makers and experts from different functional areas within the organization should be involved. So with considering the defined problem scope and its entire dimension, we must form a

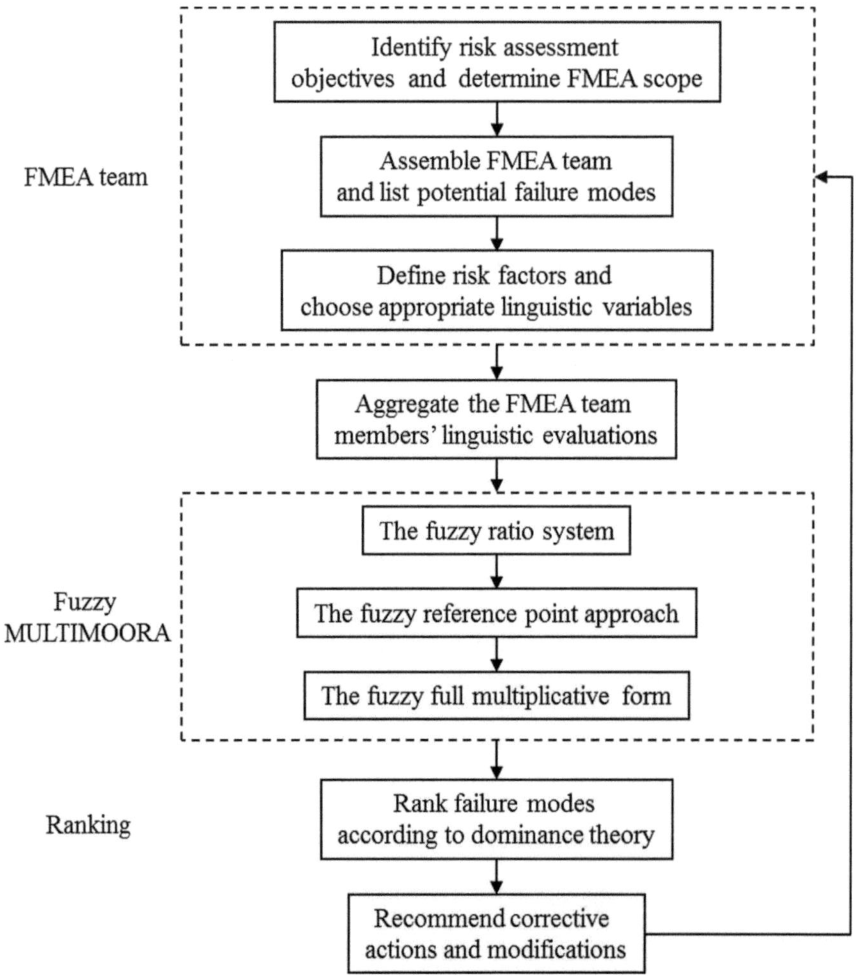

Fig. 11.1 Flowchart of the proposed FMEA approach (Liu et al. 2014)

team of cross-functional experts with one person responsible for coordinating the entire FMEA process. Then, the team should review a blueprint of the product or a detailed flowchart of the operation to understand the product or process to be studied. Basic tools such as brainstorming sessions and cause–effect diagrams can be employed to list all potential failure modes, the causes leading to them and their potential effects for each function that is analyzed.

Step 3. Define related risk factors and choose appropriate linguistic variables

In this step, it is required to define a finite set of risk factors and their evaluation metrics in order to assess the risk of failure modes. These risk factors must be defined according to the organization actual situations, the risk assessment objectives, the scope of risk analysis, and the type of product/process which will be

Table 11.1 Linguistic terms for rating failure modes

Linguistic terms	Fuzzy numbers
Extremely Low (EL)	(0, 0, 0, 0)
Very Low (VL)	(0, 0, 1, 2)
Low (L)	(1, 2, 2, 3)
Medium Low (ML)	(2, 3, 4, 5)
Medium (M)	(4, 5, 5, 6)
Medium High (MH)	(5, 6, 7, 8)
High (H)	(7, 8, 8, 9)
Very High (VH)	(8, 9, 10, 10)
Extremely High (EH)	(10, 10, 10, 10)

analyzed. In addition, we must choose appropriate linguistic terms for the importance weights of risk factors and the ratings of failure modes with regard to each risk factor. These linguistic terms can also be expressed in positive trapezoidal fuzzy numbers. It is recommended that in this chapter the FMEA team members use the linguistic terms shown in Tables 11.1 and 7.2 to evaluate the importance of risk factors and the ratings of failure modes regarding various risk factors.

Step 4. Aggregate the FMEA team members' linguistic evaluations

After the FMEA team members give their judgments on risk factors using linguistic terms, the aggregated fuzzy ratings of failure modes with respect to each risk factor can be calculated to construct the fuzzy group assessment matrix $\tilde{R} = (\tilde{r}_{ij})_{m \times n}$, where

$$\tilde{r}_{ij} = (r_{ij1}, r_{ij2}, r_{ij3}, r_{ij4}) = \left(\sum_{k=1}^{l} \lambda_k r_{ij1}^k, \sum_{k=1}^{l} \lambda_k r_{ij2}^k, \sum_{k=1}^{l} \lambda_k r_{ij3}^k, \sum_{k=1}^{l} \lambda_k r_{ij4}^k \right). \quad (11.6)$$

Similarly, the aggregated fuzzy weight for each risk factor $\tilde{w}_j$ is calculated as

$$\tilde{w}_j = (w_{j1}, w_{j2}, w_{j3}, w_{j4}) = \left(\sum_{k=1}^{l} \lambda_k w_{j1}^k, \sum_{k=1}^{l} \lambda_k w_{j2}^k, \sum_{k=1}^{l} \lambda_k w_{j3}^k, \sum_{k=1}^{l} \lambda_k w_{j4}^k \right). \quad (11.7)$$

Step 5. The fuzzy ratio system.

The fuzzy ratio system defines normalization of the fuzzy numbers $\tilde{r}_{ij}$ resulting in the normalized fuzzy assessment matrix $\tilde{X} = [\tilde{x}_{ij}]_{m \times n}$. The normalization is performed by comparing appropriate values of fuzzy numbers:

$$\tilde{x}_{ij} = (x_{ij1}, x_{ij2}, x_{ij3}, x_{ij4}) = \left(w_{j1} \frac{r_{ij1}}{\hat{r}}, w_{j2} \frac{r_{ij2}}{\hat{r}}, w_{j3} \frac{r_{ij3}}{\hat{r}}, w_{j4} \frac{r_{ij4}}{\hat{r}}, \right),$$
$$\hat{r}_j = \sqrt{\sum_{i=1}^{m} r_{ij4}^2}. \quad (11.8)$$

Once the normalized fuzzy assessment matrix $\tilde{X}$ is acquired, the summarizing ratio $\tilde{y}_i$ for each failure mode can be computed by using the following equation:

$$\tilde{y}_i = \sum_{j=1}^{g} \tilde{x}_{ij} - \sum_{j=g+1}^{n} \tilde{x}_{ij}, \tag{11.9}$$

where g = 1, 2,…,n stands for number of factors to be minimized. Then, each ratio is defuzzified by the centroid defuzzification method (cf. Definition 7.6) and the failure modes with higher defuzzified values $\bar{y}_i$ are attributed with higher ranks.

Step 6. The fuzzy reference point approach

The fuzzy maximal objective reference point (MORP) vector $\tilde{r}^* = \left(\tilde{x}_1^*, \tilde{x}_2^*, \ldots, \tilde{x}_n^*\right)$ is obtained according to the matrix $\tilde{X} = \left[\tilde{x}_{ij}\right]_{m \times n}$. The jth coordinate of the reference point resembles the fuzzy maximum or minimum of the jth risk factor $\tilde{x}_j^*$, i.e.,

$$\tilde{x}_j^* = \begin{cases} \left(\max_i x_{ij1}, \max_i x_{ij2}, \max_i x_{ij3}, \max_i x_{ij4}\right), & j \leq g; \\ \left(\min_i x_{ij1}, \min_i x_{ij2}, \min_i x_{ij3}, \min_i x_{ij4}\right), & j > g. \end{cases} \tag{11.10}$$

The distance of each failure mode from the fuzzy MORP can be currently calculated by using Eq. (11.1).

$$d_i = \max_j d\left(\tilde{x}_j^*, \tilde{x}_{ij}\right), \tag{11.11}$$

Then, the ranking orders of all failure modes are determined according to the deviation from the reference point and the min–max Metric of Tchebycheff.

Step 7. The fuzzy full multiplicative form

The overall utility of the ith failure mode can be expressed as dimensionless fuzzy number by

$$\tilde{U}_i = \tilde{A}_i \oslash \tilde{B}_i, \tag{11.12}$$

where $\tilde{A}_i = \prod_{j=1}^{g} \tilde{x}_{ij}$ denotes the product of factors of the ith failure mode to be minimized and $\tilde{B}_i = \prod_{j=g+1}^{n} \tilde{x}_{ij}$ is the product of factors of the ith failure mode to be maximized. Then, the overall utility $\tilde{U}_i$ is transformed into crisp values $\bar{U}_i$ by using the centroid method to rank the failure modes. The higher the $\bar{U}_i$ is, the higher the rank of certain failure mode.

Step 8. Determine the final ranking of failure modes based on the three ranking lists derived in previous steps, referring to the dominance theory.

Step 9. Analyze the results and develop recommendations to enhance the system performance.

Having obtained the ranking of failure modes, corrective actions should be taken by relevant departments beginning with the riskiest failures.

11.3 An Illustrative Example

11.3.1 Implementation

In what follows, a case study of preventing infant abduction (Chang et al. 2012; Liu et al. 2014) is provided to illustrate the practicality and usefulness of the proposed fuzzy FMEA approach. Infant abduction is a serious risk exposure for hospitals. Such a horrific event can impose monumental injury on family members, the facility, and its staff, as well as the community. Therefore, ensuring the safety of infants born in a hospital is a top priority and requires a solid infant security plan. A multi-facility healthcare system consisting of 629 acute care beds intends to conduct an FMEA project to minimize the potential for infant abduction. After developing a flowchart of service process, 16 potential failure modes were explored and listed through brainstorming. These failure modes, the risk factors (*O*, *S*, and *D*), and the calculated RPN for each failure mode are presented in Table 11.2. A FMEA team of five medical experts, $\text{TM}_k (k = 1, 2, \ldots, 5)$, has been formed to

Table 11.2 FMEA of the infant abduction (Liu et al. 2014)

No.	Failure modes	O	S	D	RPN
FM1	Child not banded	7	10	5	350
FM2	Insufficient IS info provided to mom	4	5	8	160
FM3	Mom not paying attention	8	5	8	320
FM4	Info not understood	2	5	8	80
FM5	Baby may not be HUGS banded prior to washing	9	10	3	270
FM6	Info not entered into computer system	8	10	5	400
FM7	Delay in entering info into computer system	4	10	5	200
FM8	Unfounded alarms	3	10	10	300
FM9	Alarm ringing—doors not locking	2	10	10	200
FM10	HUGS band not applied until reaching post partum	5	10	2	100
FM11	Bands loosening	9	8	6	432
FM12	Bands not checked and/or tightened properly	3	8	8	192
FM13	Not checked against census	8	7	7	392
FM14	Transferred rooms, not updated	7	7	7	343
FM15	HUGS band may not be checked when moving to nursery	7	5	3	105
FM16	Leaving SCN other than for discharge without HUGS band	5	8	8	320

conduct the risk evaluation and to identify the most serious failure modes for taking preventive measures. The five team members from different departments are assigned the following relative weights: 0.15, 0.20, 0.30, 0.20, and 0.15 in the risk analysis process.

Table 11.3 summarizes the linguistic evaluation information about the sixteen failure modes with respect to the risk factors *O*, *S*, and *D*. The risk factors and their relative weights are expressed by using the linguistic variables given in Tables 11.1 and 7.2. To be specific, the three risk factors are expressed in a nine-point scale, whereas the relative importance of risk factors is mapped onto a seven-point scale. Next, the fuzzy assessment matrix from experts will be analyzed by applying the proposed fuzzy FMEA to identify the most critical failure modes.

After translating into corresponding fuzzy numbers, the FMEA team members' linguistic evaluations are aggregated using Eqs. (11.6) and (11.7) to construct the fuzzy group assessment matrix $\tilde{R} = \left(\tilde{r}_{ij}\right)_{16\times 3}$ and to get the aggregated fuzzy weights of risk factors. The results are shown in Table 11.4.

Firstly, the ranking of failure modes is performed in accordance with the fuzzy ratio system. The aggregated fuzzy assessment matrix is normalized by employing Eq. (11.8) and the normalized fuzzy assessment matrix $\tilde{X} = \left[\tilde{x}_{ij}\right]_{16\times 3}$ is presented in Table 11.5. Then, the normalized data are aggregated by using Eq. (11.9) and defuzzified according to the centroid method. The sixteen failures are then ranked in decreasing order of the crisp values as reported in Table 11.6.

Secondly, the fuzzy MORP is defined according to Eq. (11.10), as shown in the last row of Table 11.5. The distances from fuzzy MORP are then calculated by employing Eq. (11.11) for all the identified failure modes and the results are shown in Table 11.7. The failure modes are ranked in ascending order of the maximal deviations.

Thirdly, the failure modes are ranked according to the fuzzy multiplicative form as described by Eq. (11.12). Given large numbers involved in the computing, Table 11.8 presents the summarized data only.

Finally, the theory of dominance is employed to aggregate the three rank lists provided by different parts of the fuzzy MULTIMOORA into a single final rank. The last column in Table 11.9 presents the final ranking of the failure modes identified in the FMEA. Accordingly, the risk priority ranking of failure modes is FM11 $\succ$ FM13 $\succ$FM6 $\succ$ FM1 $\succ \cdots \succ$ FM10 $\succ$ FM4 in terms of the risk factors, *O*, *S*, and *D*, by the proposed FMEA model. Thus, FM11 is determined as the most serious failure mode and should be given the top risk priority by the medical center; this will be followed by FM13, FM6, FM1, FM16, FM3, …, FM10, and FM4. The obtained results in ranking of potential failure modes can provide for risk decision-making support in developing corrective actions to protect against infant or child abduction in the healthcare facility.

Table 11.3 Linguistic evaluations on failure modes by the five FMEA team members (Liu et al. 2014)

Risk factors		*O*					*S*					*D*				
Team members		TM1	TM2	TM3	TM4	TM5	TM1	TM2	TM3	TM4	TM5	TM1	TM2	TM3	TM4	TM5
Failure modes	FM1	MH	H	MH	H	MH	EH	VH	EH	EH	EH	M	M	M	M	M
	FM2	ML	ML	M	ML	ML	M	M	M	M	M	H	MH	H	H	H
	FM3	H	H	H	VH	MH	M	MH	M	M	ML	MH	H	H	H	MH
	FM4	ML	L	L	L	L	M	M	M	ML	M	H	MH	H	H	H
	FM5	VH	H	VH	VH	VH	EH	EH	EH	VH	EH	ML	L	L	ML	ML
	FM6	H	H	H	H	H	EH	EH	EH	EH	EH	M	ML	M	MH	M
	FM7	M	ML	ML	M	ML	VH	EH	EH	EH	EH	M	M	M	M	M
	FM8	ML	ML	L	ML	L	EH	EH	EH	EH	EH	EH	EH	EH	EH	EH
	FM9	L	L	L	L	L	EH	EH	VH	EH	EH	EH	EH	VH	EH	EH
	FM10	M	M	M	ML	MH	EH	EH	EH	EH	VH	L	L	VL	L	ML
	FM11	H	VH	VH	H	VH	H	H	VH	H	MH	M	M	MH	MH	MH
	FM12	ML	L	ML	ML	ML	MH	H	MH	H	H	H	H	MH	H	VH
	FM13	H	MH	VH	H	H	MH	MH	MH	MH	H	MH	MH	H	MH	MH
	FM14	MH	H	MH	MH	MH	H	MH	M	MH	MH	M	MH	MH	M	MH
	FM15	MH	MH	MH	MH	H	M	M	MH	M	M	L	ML	ML	L	L
	FM16	H	MH	H	MH	MH	VH	EH	EH	EH	EH	M	ML	M	M	M
Weights		H	VH	H	VH	H	VH	VH	VH	VH	VH	MH	H	H	H	VH

Table 11.4 Fuzzy group assessment matrix and aggregated fuzzy weights of risk factors (Liu et al. 2014)

Failure modes	*O*	*S*	*D*
FM1	(5.8, 6.8, 7.4, 8.4)	(9.6, 9.8, 10, 10)	(4, 5, 5, 6)
FM2	(2.6, 3.6, 4.3, 5.3)	(4, 5, 5, 6)	(6.6, 7.6, 7.8, 8.8)
FM3	(6.9, 7.9, 8.25, 9.05)	(3.9, 4.9, 5.25, 6.25)	(6.4, 7.4, 7.7, 8.7)
FM4	(1.15, 2.15, 2.3, 3.3)	(3.6, 4.6, 4.8, 5.8)	(6.6, 7.6, 7.8, 8.8)
FM5	(7.8, 8.8, 9.6, 9.8)	(9.6, 9.8, 10, 10)	(1.5, 2.5, 3, 4)
FM6	(7, 8, 8, 9)	(10, 10, 10, 10)	(3.8, 4.8, 5.2, 6.2)
FM7	(2.7, 3.7, 4.35, 5.35)	(9.7, 9.85, 10, 10)	(4, 5, 5, 6)
FM8	(1.55, 2.55, 3.1, 4.1)	(10, 10, 10, 10)	(10, 10, 10, 10)
FM9	(1, 2, 2, 3)	(9.4, 9.7, 10, 10)	(9.4, 9.7, 10, 10)
FM10	(3.75, 4.75, 5.1, 6.1)	(9.7, 9.85, 10, 10)	(0.85, 1.55, 2, 3)
FM11	(7.65, 8.65, 9.3, 9.65)	(7, 8, 8.45, 9.15)	(4.65, 5.65, 6.3, 7.3)
FM12	(1.8, 2.8, 3.6, 4.6)	(6.1, 7.1, 7.55, 8.55)	(6.55, 7.55, 8, 8.85)
FM13	(6.9, 7.9, 8.4, 9.1)	(5.3, 6.3, 7.15, 8.15)	(5.6, 6.6, 7.3, 8.3)
FM14	(5.4, 6.4, 7.2, 8.2)	(5, 6, 6.55, 7.55)	(4.65, 5.65, 6.3, 7.3)
FM15	(5.3, 6.3, 7.15, 8.15)	(4.3, 5.3, 5.6, 6.6)	(1.5, 2.5, 3, 4)
FM16	(5.9, 6.9, 7.45, 8.45)	(9.7, 9.85, 10, 10)	(3.6, 4.6, 4.8, 5.8)
Aggregated weights	(0.74, 0.84, 0.88, 0.94)	(0.8, 0.9, 1, 1)	(0.685, 0.785, 0.815, 0.9)

11.3.2 Sensitivity Analysis

A sensitivity analysis by changing the weights of risk factors is conducted according to the information given in Table 11.4. The ranking results for the sixteen failure modes with respect to different cases are represented in Table 11.10. Note that Case 0 shows the original weights of the risk factors while the other cases show different risk factor weights for possible situations.

As one can see, FM11 is the failure mode with the top risk priority in three of the four cases. In Case 0, FM13 is the second most important failure mode where the weight of S is relatively high, whereas the weights of *O* and *D* are relatively low. In Case 1 and Case 2, FM6 is at the second position since the weight of *D* is relatively low. As the weight of *D* is the highest, FM9 becomes the second most important failure mode in Case 5. The sensitivity analysis indicates that the weights of risk factors can have a great influence on the final ranking orders of failure modes. Therefore, in real-world scenarios, determining suitable risk factor weights according to actual situations and experts' opinions is of significance and benefit to the risk prioritization of failure modes and the following corrective actions.

Table 11.5 Normalized fuzzy assessment matrix and fuzzy MORP vector (Liu et al. 2014)

Failure modes	*O*	*S*	*D*
FM1	(0.146, 0.195, 0.222, 0.269)	(0.219, 0.251, 0.285, 0.285)	(0.093, 0.133, 0.138, 0.183)
FM2	(0.066, 0.103, 0.129, 0.170)	(0.091, 0.128, 0.142, 0.171)	(0.153, 0.202, 0.216, 0.269)
FM3	(0.174, 0.226, 0.247, 0.290)	(0.089, 0.126, 0.150, 0.178)	(0.149, 0.197, 0.213, 0.266)
FM4	(0.029, 0.062, 0.069, 0.106)	(0.082, 0.118, 0.137, 0.165)	(0.153, 0.202, 0.216, 0.269)
FM5	(0.197, 0.252, 0.288, 0.314)	(0.219, 0.251, 0.285, 0.285)	(0.035, 0.067, 0.083, 0.122)
FM6	(0.176, 0.229, 0.240, 0.288)	(0.228, 0.256, 0.285, 0.285)	(0.088, 0.128, 0.144, 0.189)
FM7	(0.068, 0.106, 0.130, 0.171)	(0.221, 0.253, 0.285, 0.285)	(0.093, 0.133, 0.138, 0.183)
FM8	(0.039, 0.073, 0.093, 0.131)	(0.228, 0.256, 0.285, 0.285)	(0.232, 0.266, 0.276, 0.305)
FM9	(0.025, 0.057, 0.060, 0.096)	(0.214, 0.249, 0.285, 0.285)	(0.218, 0.258, 0.276, 0.305)
FM10	(0.095, 0.136, 0.153, 0.195)	(0.221, 0.253, 0.285, 0.285)	(0.020, 0.041, 0.055, 0.092)
FM11	(0.193, 0.247, 0.279, 0.309)	(0.160, 0.205, 0.241, 0.261)	(0.108, 0.150, 0.174, 0.223)
FM12	(0.045, 0.080, 0.108, 0.147)	(0.139, 0.182, 0.215, 0.244)	(0.152, 0.201, 0.221, 0.270)
FM13	(0.174, 0.226, 0.252, 0.291)	(0.121, 0.162, 0.204, 0.232)	(0.130, 0.176, 0.202, 0.253)
FM14	(0.136, 0.183, 0.216, 0.262)	(0.114, 0.154, 0.187, 0.215)	(0.108, 0.150, 0.174, 0.223)
FM15	(0.134, 0.180, 0.214, 0.261)	(0.098, 0.136, 0.160, 0.188)	(0.035, 0.067, 0.083, 0.122)
FM16	(0.149, 0.197, 0.223, 0.270)	(0.221, 0.253, 0.285, 0.285)	(0.084, 0.122, 0.133, 0.177)
$\tilde{x}^*$	(0.197, 0.252, 0.288, 0.314)	(0.228, 0.256, 0.285, 0.285)	(0.232, 0.266, 0.276, 0.305)

11.3.3 Comparisons and Discussion

To further illustrate the effectiveness of the proposed FMEA model, we used the above case study to analyze some comparable methods, which include the conventional RPN and the crisp MULTIMOORA. Figure 11.2 exhibits the ranking results of all the sixteen failure modes as obtained using these approaches. It is clearly shown in Fig. 11.2 that most of the failure modes have the same rank orders in the three different FMEA approaches. The Spearman's rank-correlation coefficients between the risk ranking lists by the proposed method and the conventional RPN and the crisp MULTIMOORA methods are 0.982 and 0.985, respectively. This demonstrates the validity of the presented fuzzy FMEA. However, there are also some differences between the ranking orders obtained by the three approaches. These inconsistent ranking results can be in part explained by the limitations of the conventional RPN and the crisp MULTIMOORA methods. For example, both FM3 and FM16 have the same RPN = 320. And both FM7 and FM9 have the same RPN = 200. That is, the failure modes with different combinations of *O*, *S*, and *D* produce the same value of RPN, leading to difficult decision making by the

Table 11.6 The fuzzy ratio system (Liu et al. 2014)

Failure modes	$\tilde{y}_i$	$\bar{y}_i$	Ranking
FM1	(0.458, 0.579, 0.645, 0.737)	0.6031	5
FM2	(0.310, 0.433, 0.487, 0.609)	0.4598	13
FM3	(0.411, 0.549, 0.610, 0.733)	0.575	9
FM4	(0.264, 0.382, 0.421, 0.539)	0.4017	16
FM5	(0.450, 0.570, 0.655, 0.721)	0.5966	7
FM6	(0.493, 0.613, 0.668, 0.762)	0.6326	2
FM7	(0.382, 0.491, 0.553, 0.639)	0.5153	11
FM8	(0.499, 0.596, 0.654, 0.721)	0.6161	3
FM9	(0.458, 0.564, 0.621, 0.686)	0.5802	8
FM10	(0.335, 0.430, 0.493, 0.572)	0.4566	14
FM11	(0.460, 0.603, 0.694, 0.792)	0.6352	1
FM12	(0.337, 0.463, 0.544, 0.661)	0.5007	12
FM13	(0.425, 0.563, 0.657, 0.777)	0.6046	4
FM14	(0.358, 0.487, 0.576, 0.700)	0.5303	10
FM15	(0.266, 0.383, 0.457, 0.571)	0.4191	15
FM16	(0.453, 0.572, 0.641, 0.732)	0.5983	6

Table 11.7 The fuzzy reference point approach (Liu et al. 2014)

Failure modes	O	S	D	d_i	Ranking
FM1	0.0574	0.0048	0.1341	0.1341	6
FM2	0.1487	0.1324	0.0621	0.1487	9
FM3	0.0308	0.1301	0.0661	0.1301	4
FM4	0.1998	0.1401	0.0621	0.1998	14
FM5	0.0000	0.0048	0.1946	0.1946	12
FM6	0.0334	0.0000	0.1340	0.1340	5
FM7	0.1466	0.0036	0.1341	0.1466	8
FM8	0.1816	0.0000	0.0000	0.1816	11
FM9	0.2066	0.0071	0.0073	0.2066	15
FM10	0.1209	0.0036	0.2198	0.2198	16
FM11	0.0063	0.0490	0.1080	0.1080	2
FM12	0.1704	0.0712	0.0609	0.1704	10
FM13	0.0287	0.0870	0.0823	0.0870	1
FM14	0.0659	0.0985	0.1080	0.1080	2
FM15	0.0680	0.1202	0.1946	0.1946	12
FM16	0.0553	0.0036	0.1421	0.1421	7

Table 11.8 The fuzzy full multiplicative form (Liu et al. 2014)

Failure modes	$\tilde{U}_i$	$\bar{U}_i$	Ranking
FM1	(0.0030, 0.0065, 0.0087, 0.0140)	0.00816	4
FM2	(0.0009, 0.0027, 0.0040, 0.0078)	0.00395	13
FM3	(0.0023, 0.0056, 0.0079, 0.0137)	0.00750	6
FM4	(0.0004, 0.0015, 0.0020, 0.0047)	0.00224	16
FM5	(0.0015, 0.0042, 0.0068, 0.0109)	0.00592	9
FM6	(0.0035, 0.0075, 0.0098, 0.0155)	0.00920	3
FM7	(0.0014, 0.0036, 0.0051, 0.0089)	0.00484	10
FM8	(0.0021, 0.0050, 0.0073, 0.0114)	0.00650	8
FM9	(0.0012, 0.0037, 0.0047, 0.0083)	0.00455	12
FM10	(0.0004, 0.0014, 0.0024, 0.0051)	0.00242	15
FM11	(0.0033, 0.0076, 0.0117, 0.0179)	0.01024	1
FM12	(0.0010, 0.0029, 0.0051, 0.0097)	0.00481	11
FM13	(0.0027, 0.0064, 0.0103, 0.0171)	0.00930	2
FM14	(0.0017, 0.0042, 0.0070, 0.0126)	0.00652	7
FM15	(0.0005, 0.0016, 0.0028, 0.0060)	0.00283	14
FM16	(0.0027, 0.0061, 0.0084, 0.0136)	0.00783	5

Table 11.9 Final ranking of failure modes by the proposed fuzzy FMEA (Liu et al. 2014)

Failure modes	The fuzzy ratio system	The fuzzy reference point	The fuzzy full multiplicative form	Final ranking
FM1	5	6	4	4
FM2	13	9	13	13
FM3	9	4	6	6
FM4	16	14	16	16
FM5	7	12	9	9
FM6	2	5	3	3
FM7	11	8	10	10
FM8	3	11	8	8
FM9	8	15	12	12
FM10	14	16	15	15
FM11	1	2	1	1
FM12	12	10	11	11
FM13	4	1	2	2
FM14	10	2	7	7
FM15	15	12	14	14
FM16	6	7	5	5

Table 11.10 Risk priority rankings with respect to the considered cases (Liu et al. 2014)

Failure modes	Case 0	Case 1	Case 2	Case 3
	$w_O = 0.3$, $w_S = 0.4$, $w_D = 0.3$	$w_O = 0.6$, $w_S = 0.2$, $w_D = 0.2$	$w_O = 0.2$, $w_S = 0.6$, $w_D = 0.2$	$w_O = 0.2$, $w_S = 0.2$, $w_D = 0.6$
FM1	4	5	3	9
FM2	13	15	14	8
FM3	6	4	13	3
FM4	16	16	16	12
FM5	9	8	6	14
FM6	3	2	2	7
FM7	10	11	7	13
FM8	8	10	5	1
FM9	12	13	8	2
FM10	15	14	11	16
FM11	1	1	1	5
FM12	11	12	10	6
FM13	2	3	9	4
FM14	7	7	12	10
FM15	14	9	15	15
FM16	5	6	4	11

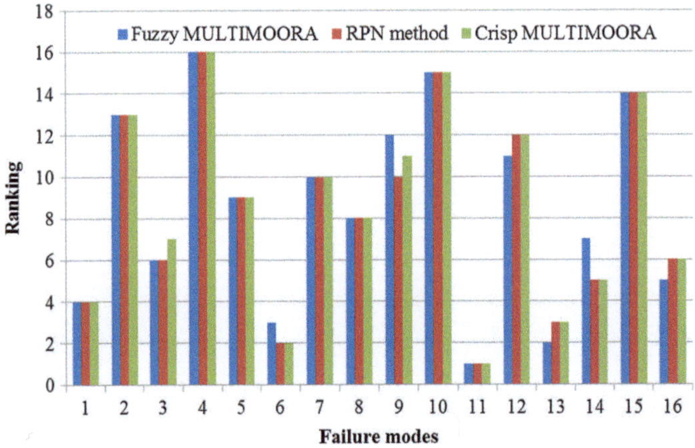

Fig. 11.2 Comparative ranking of failure modes for the considered example (Liu et al. 2014)

traditional FMEA. However, this problem can be easily solved by using the MULTIMOORA method. The results of the proposed model and the crisp MULTIMOORA show that for FM16 and FM7, more urgently corrective actions are needed. In addition, the ranking orders of FM6, FM12, FM13, and FM14 are different from the ones produced by the crisp MULTIMOORA which are, however, in agreement with the ranking results of the conventional RPN. This is mainly because the imprecise and uncertain information is not considered in the conventional RPN and the crisp MULTIMOORA, thus causing biased ranking results.

The comparison analysis shows that a more accurate and reasonable ranking can be determined by the application of fuzzy set theory and MULTIMOORA method to FMEA. The proposed model is superior to other risk analysis methods since it has capability of representing the vague knowledge and expertise of FMEA team members. In risk evaluation problems, data are very often imprecise and fuzzy. Risk analysts may encounter difficulty in quantifying such data. The fuzzy FMEA model proposed in this chapter easily quantifies these types of data. Moreover, the MULTIMOORA method was employed in order to determine the risk priority of failure modes and thus identify the high-risk failure modes. It includes an effective method to weight the risk factors and to rank the identified failure modes. Therefore, the proposed FMEA model might be suitable when conducting risk analyses which require quantitative as well as qualitative inputs.

References

Brauers WKM, Zavadskas EK (2006) The MOORA method and its application to privatization in a transition economy. Control Cybernet 35(2):445–469

Brauers WKM, Zavadskas EK (2010) Project management by MULTIMOORA as an instrument for transition economies. Technol Econ Dev Econ 16(1):5–24

Brauers WKM, Zavadskas EK (2011) MULTIMOORA optimization used to decide on a bank loan to buy property. Technol Econ Dev Econ 17(1):174

Brauers WKM, Zavadskas EK (2012) Robustness of MULTIMOORA: a method for multi-objective optimization. Informatica 23(1):1–25

Chang DS, Chung JH, Sun KLP, Yang FC (2012) A novel approach for evaluating the risk of health care failure modes. J Med Syst 36(6):3967–3974

Liu HC, Fan XJ, Li P, Chen YZ (2014) Evaluating the risk of failure modes with extended MULTIMOORA method under fuzzy environment. Eng Appl Artif Intell 34:168–177

Wan SP, Li DF (2013) Fuzzy LINMAP approach to heterogeneous MADM considering comparisons of alternatives with hesitation degrees. Omega 41(6):925–940

Part V
FMEA Based on Hybrid MCDM Methods

Chapter 12
FMEA Using Combination Weighting and Fuzzy VIKOR Method

Due to its characteristics and capabilities, the VIKOR method has been employed by Liu et al. (2012) to resolve the risk evaluation problem under fuzzy environment. To overcome the shortcomings and enhance the assessment capability of FMEA, Liu et al. (2015) further presented a hybrid MCDM approach for risk analysis based on combination weighting and fuzzy VIKOR method. Combination of fuzzy analytic hierarchy process (AHP) and entropy method is applied for risk factor weighting in the proposed approach. The risk priorities of the identified failure modes are obtained through next steps based on fuzzy VIKOR method. The combination weighting method considering both subjective and objective weights of risk factors is helpful to reflect the essential characteristics of the risk evaluation problem. In addition, the fuzzy VIKOR method helps decision makers in FMEA achieve an acceptable compromise of the maximum group utility for the "majority" and the minimum of the individual regret for the "opponent."

12.1 Preliminaries

12.1.1 Fuzzy Set Theory

The basic definitions related to fuzzy sets and triangular fuzzy numbers can be found in Sects. 7.1.1 and 11.1.1. In this section, we only introduce the distance between triangular fuzzy numbers and the center of area (COA) method which will be utilized in the proposed FMEA model.

Definition 12.1 According to Chen (2000), the distance between the triangular fuzzy numbers $\tilde{a} = (a_1, a_2, a_3)$ and $\tilde{b} = (b_1, b_2, b_3)$ is calculated by using the vertex method as

H.-C. Liu, *FMEA Using Uncertainty Theories and MCDM Methods*,
DOI 10.1007/978-981-10-1466-6_12

$$d(\tilde{a},\tilde{b}) = \sqrt{\frac{1}{3}\left[(a_1 - b_1)^2 + (a_2 - b_2)^2 + (a_3 - b_3)^2\right]}. \quad (12.1)$$

Definition 12.2 Using the COA method, the crisp value of the triangular fuzzy number $\tilde{a} = (a_1, a_2, a_3)$ is expressed by the following relation (Liu et al. 2015):

$$\bar{x}_0(\tilde{a}) = \frac{1}{3}[(a_3 - a_1) + (a_2 - a_1)] + a_1, \quad (12.2)$$

where $\bar{x}_0(\tilde{a})$ is the defuzzified value of the fuzzy number $\tilde{a}$.

12.1.2 Fuzzy AHP Method

The analytic hierarchy process (AHP) (Saaty 1980) is a useful approach to tackle the complexity of decision problems by means of a hierarchy of decision layers. However, the classical AHP uses exact numerical values in the pairwise comparison matrix and is not fully capable of reflecting the human judgments. As a result, fuzzy extension of the AHP (Buckley et al. 2001) was presented to ease its adaptation to real-life problems, which is employed in this chapter to calculate subjective risk factor weights.

Assuming l decision makers DM$_k$ $(k = 1, 2, \ldots, l)$, we proceed to make decision on m alternatives with n criteria. Each decision maker DM$_k$ is given a weight $\lambda_k > 0\,(k = 1, \ldots, l)$ satisfying $\sum_1^l \lambda_k = 1$ to reflect his/her relative importance in the decision-making process. The procedure for determining the weights of criteria by using the fuzzy AHP method is summarized as follows (Liu et al. 2015):

Step 1. Compare the performance score

Through expert questionnaires, each expert is asked to assign linguistic terms expressed by triangular fuzzy numbers (see Table 12.1 and Fig. 12.1) to the pairwise comparisons among all criteria in the dimensions of a hierarchy system. Let $\tilde{a}_{ij}^k = \left(a_{ij1}^k, a_{ij2}^k, a_{ij3}^k\right)$ $(i = 1, 2, \ldots, (n-1), j = 2, 3, \ldots, n)$ be the fuzzy relative importance by comparing criterion i with criterion j provided by the kth decision

Table 12.1 Linguistic terms for rating risk factor weights

Fuzzy numbers	Linguistic terms	Triangular fuzzy numbers
$\tilde{9}$	Absolutely important (AI)	(7, 9, 9)
$\tilde{7}$	Very strongly important (VSI)	(5, 7, 9)
$\tilde{5}$	Strong important (SI)	(3, 5, 7)
$\tilde{3}$	Weakly important (WI)	(1, 3, 5)
$\tilde{1}$	Equally important (EI)	(1, 1, 3)

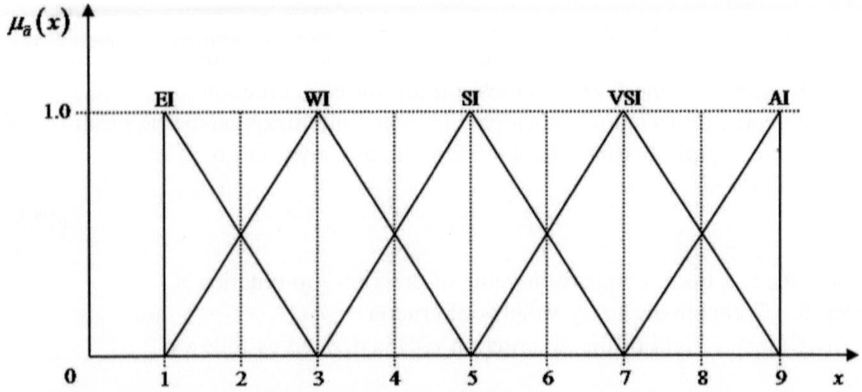

Fig. 12.1 Membership functions for rating risk factor weights

maker. Then, the aggregated fuzzy relative importance ($\tilde{a}_{ij}$) can be calculated as follows:

$$\tilde{a}_{ij} = \left(a_{ij1}, a_{ij2}, a_{ij3}\right), \quad i = 1, 2, \ldots, n-1, \quad j = 2, 3, \ldots, n, \tag{12.3}$$

where

$$a_{ij1} = \sum_{k=1}^{l} \lambda_k a_{ij1}^k, \quad a_{ij2} = \sum_{k=1}^{l} \lambda_k a_{ij2}^k, \quad a_{ij3} = \sum_{k=1}^{l} \lambda_k a_{ij3}^k.$$

Step 2. Construct the fuzzy pairwise comparison matrix

The result of the comparisons is constructed as a fuzzy pairwise comparison matrix ($\tilde{A}$), such that

$$\tilde{A} = \left[\tilde{a}_{ij}\right] = \begin{bmatrix} \tilde{a}_{11} & \tilde{a}_{12} & \ldots & \tilde{a}_{1n} \\ \tilde{a}_{21} & \tilde{a}_{22} & \ldots & \tilde{a}_{2n} \\ \vdots & \vdots & \ldots & \vdots \\ \tilde{a}_{n1} & \tilde{a}_{n2} & \ldots & \tilde{a}_{nn} \end{bmatrix} = \begin{bmatrix} 1 & \tilde{a}_{12} & \ldots & \tilde{a}_{1n} \\ 1/\tilde{a}_{12} & 1 & \ldots & \tilde{a}_{2n} \\ \vdots & \vdots & \ldots & \vdots \\ 1/\tilde{a}_{1n} & 1/\tilde{a}_{2n} & \ldots & 1 \end{bmatrix}. \tag{12.4}$$

Step 3. Examine consistency of the fuzzy pairwise comparison matrix

Assume $A = \left[a_{ij}\right]$ is a positive reciprocal matrix and $\tilde{A} = \left[\tilde{a}_{ij}\right]$ is a fuzzy positive reciprocal matrix. As pointed out by Buckley et al. (2001), if $A = \left[a_{ij}\right]$ is consistent,

$\tilde{A} = [\tilde{a}_{ij}]$ will also be consistent. In case the consistency of the comparison matrix is not verified, the evaluation procedure has to be repeated to improve consistency.

Step 4. Compute the fuzzy geometric mean for each criterion

The geometric technique is adopted to define the fuzzy geometric mean ($\tilde{r}_i$) of the fuzzy comparison values between criteria, as shown in Eq. (12.5).

$$\tilde{r}_i = (\tilde{a}_{i1} \otimes \tilde{a}_{i2} \otimes \cdots \otimes \tilde{a}_{in})^{1/n}, \tag{12.5}$$

where $\tilde{a}_{in}$ is a fuzzy comparison value of criterion i to criterion n.

Step 5. Compute the fuzzy weights of criteria

The fuzzy weight of the ith criterion ($\tilde{w}_i^s$) is derived as follows:

$$\tilde{w}_i^s = \tilde{r}_i \otimes (\tilde{r}_1 \oplus \tilde{r}_2 \oplus \cdots \oplus \tilde{r}_n)^{-1}, \tag{12.6}$$

where $\tilde{w}_i^s$ can be indicated by a triangular fuzzy number, $\tilde{w}_i^s = (w_{i1}^s, w_{i2}^s, w_{i3}^s)$.

Step 6. Defuzzify the values of $\tilde{w}_i^s$

The subjective weight of criterion i (w_i^s) is first defuzzified using Eq. (12.2) and then normalized by

$$w_i^s = \frac{\bar{w}_i^s}{\sum_{i=1}^{n} \bar{w}_i^s}, \tag{12.7}$$

where $\bar{w}_i^s$ is referred to as the crisp number of the fuzzy weight $\tilde{w}_i^s$.

12.1.3 *Shannon Entropy*

Shannon entropy (Shannon and Weaver 1947) is a measure of information uncertainty formulated in terms of probability theory. It is well suited for measuring the relative contrast intensities of criteria to represent the average intrinsic information transmitted to the decision maker. According to the entropy method, if all alternatives are the same in relation to a specific criterion, then that criterion should be eliminated because it transmits no information about decision makers' preferences. On the opposite, the criterion that transmits the most information should have the greatest importance weighting.

Entropy concept is capable of being deployed as an objective weighting calculation method through the following steps (Liu et al. 2015):

Step 1. Normalize the evaluation criterion as

$$P_{ij} = \frac{x_{ij}}{\sum_{i=1}^{m} x_{ij}}, \tag{12.8}$$

where P_{ij} means the projected outcomes of criterion j.

Step 2. Calculate the entropy E_j of the set of projected outcomes of criterion j using the following equation:

$$E_j = -\left(\frac{1}{\ln m}\right) \sum_{i=1}^{m} P_{ij} \ln P_{ij}, \tag{12.9}$$

where m is the number of criteria and guarantees that E_j lies between 0 and 1.

Step 3. Define the divergence through

$$div_j = 1 - E_j, \tag{12.10}$$

where div_j is the divergence degree of the intrinsic information of criterion j. The greater the value of the div_j is, the more important the criterion is in the decision-making process.

Step 4. Obtain the objective weights of criteria as follows:

$$w_j^o = \frac{div_j}{\sum_{j=1}^{n} div_j}. \tag{12.11}$$

12.1.4 Fuzzy VIKOR Method

The VIKOR method was first proposed by Opricovic (1998) for multi-criteria optimization of complex systems, which can determine compromise solutions for a problem with conflicting criteria and help the decision makers to reach a final decision. In Liu et al. (2012), a modified fuzzy approach to the normal VIKOR method was presented to process uncertain data and solve fuzzy multi-criteria problems with conflicting and non-commensurable criteria.

Suppose that a group MCDM problem has l decision makers $DM_k\,(k = 1, 2, \ldots, l)$, m alternatives $A_i\,(i = 1, 2, \ldots, m)$, and n decision criteria $C_j\,(j = 1, 2, \ldots, n)$. Each alternative is assessed with respect to the n criteria. Let $\tilde{x}_{ij}^k = \left(x_{ij1}^k, x_{ij2}^k, x_{ij3}^k\right)$ be the fuzzy rating of the ith alternative on the jth criterion provided by the kth decision maker, and let $\lambda_k\,(k = 1, \ldots, l)$ be the relative importance weights of the l decision makers, satisfying $\sum_1^l \lambda_k = 1$ and $\lambda_k > 0$ for k = 1, 2,…, l. Then, the procedure of the modified fuzzy VIKOR method consists of the following steps (Liu et al. 2015):

Step 1. Aggregate the decision makers' opinions to get the aggregated fuzzy ratings of alternatives and construct a fuzzy decision matrix

The aggregated fuzzy ratings ($\tilde{x}_{ij}$) of alternatives with respect to each criterion are calculated as follows:

$$\tilde{x}_{ij} = \left(x_{ij1}, x_{ij2}, x_{ij3}\right), \tag{12.12}$$

where

$$x_{ij1} = \sum_{k=1}^{l} \lambda_k x_{ij1}^k, \quad x_{ij2} = \sum_{k=1}^{l} \lambda_k x_{ij2}^k, \quad x_{ij3} = \sum_{k=1}^{l} \lambda_k x_{ij3}^k.$$

A group MCDM problem can be concisely expressed in matrix format as follows:

$$\tilde{D} = \begin{bmatrix} \tilde{x}_{11} & \tilde{x}_{12} & \cdots & \tilde{x}_{1n} \\ \tilde{x}_{21} & \tilde{x}_{22} & \cdots & \tilde{x}_{2n} \\ \vdots & \vdots & \cdots & \vdots \\ \tilde{x}_{m1} & \tilde{x}_{m2} & \cdots & \tilde{x}_{mn} \end{bmatrix},$$

where $\tilde{x}_{ij}$ denotes the rating of alternative A_i with respect to criterion C_j.

Step 2. Determine the fuzzy best $\tilde{f}_j^*$ and the fuzzy worst $\tilde{f}_j^-$ values of all criteria ratings, j = 1, 2,…, n

$$\tilde{f}_j^* = \left\{ \begin{array}{lll} \max\limits_i \tilde{x}_{ij}, & \text{for} & \text{benefit criteria} \\ \min\limits_i \tilde{x}_{ij}, & \text{for} & \text{cost criteria} \end{array} \right\}, \quad j = 1, 2, \ldots, n, \tag{12.13}$$

$$\tilde{f}_j^- = \left\{ \begin{array}{lll} \min\limits_i \tilde{x}_{ij}, & \text{for} & \text{benefit criteria} \\ \max\limits_i \tilde{x}_{ij}, & \text{for} & \text{cost criteria} \end{array} \right\}, \quad j = 1, 2, \ldots, n. \tag{12.14}$$

Step 3. Calculate the normalized fuzzy distance d_{ij}, i = 1, 2,…, m, j = 1, 2,…, n,

$$d_{ij} = \frac{d\left(\tilde{f}_j^*, \tilde{x}_{ij}\right)}{d\left(\tilde{f}_j^*, \tilde{f}_j^-\right)}. \tag{12.15}$$

Step 4. Compute the values S_i and R_i, i = 1, 2,…, m, by the relations

$$\begin{aligned} S_i &= \varphi \sum_{j=1}^{n} w_j^s d_{ij} + (1-\varphi) \sum_{j=1}^{n} w_j^o d_{ij} \\ &= \sum_{j=1}^{n} \left[\varphi w_j^s + (1-\varphi) w_j^o\right] d_{ij} = \sum_{j=1}^{n} w_j^c d_{ij}, \end{aligned} \tag{12.16}$$

$$R_i = \max_j \left[\varphi w_j^s d_{ij} + (1-\varphi) w_j^o d_{ij}\right] = \max_j \left(w_j^c d_{ij}\right). \tag{12.17}$$

where $w_j^c = \varphi w_j^s + (1-\varphi) w_j^o$ are the combination weights of criteria, and $\varphi \in [0,\ 1]$, expressing the relative importance between the subjective weight and the objective weight. In this chapter, the two kinds of weights are assumed to be equally important, that is, $\varphi = 0.5$.

Step 5. Compute the values Q_i, i = 1, 2,…, m, by the relation

$$Q_i = v \frac{S_i - S^*}{S^- - S^*} + (1-v) \frac{R_i - R^*}{R^- - R^*}, \tag{12.18}$$

where $S^* = \min_i S_i, S^- = \max_i S_i$, $R^* = \min_i R_i, R^- = \max_i R_i$ and v is introduced as a weight for the strategy of maximum group utility, whereas $1 - v$ is the weight of the individual regret. The value of v is set to 0.5 in this study.

Step 6. Rank the alternatives, sorting by the values S, R, and Q in decreasing order. The results are three ranking lists

Step 7. Propose a compromise solution, the alternative ($A^{(1)}$), which is the best ranked by the measure Q (minimum) if the following two conditions are satisfied:

C1. Acceptable advantage: $Q(A^{(2)}) - Q(A^{(1)}) \geq 1/(m-1)$, where $A^{(2)}$ is the alternative with the second position in the ranking list by Q.

C2. Acceptable stability in decision making: The alternative $A^{(1)}$ must also be the best ranked by S and/or R. This compromise solution is stable within a decision-making process, which could be "voting by majority rule" (when $v > 0.5$ is needed), or "by consensus" $v \approx 0.5$, or "with veto" ($v < 0.5$).

If one of the conditions is not satisfied, then a set of compromise solutions is proposed, which consists of the following:

- Alternatives $A^{(1)}$ and $A^{(2)}$ if only the condition C2 is not satisfied or
- Alternatives $A^{(1)}$, $A^{(2)}$,…, $A^{(M)}$ if the condition C1 is not satisfied; $A^{(M)}$ is determined by the relation $Q(A^{(M)}) - Q(A^{(1)}) < 1/(m-1)$ for maximum M.

12.2 The Proposed FMEA Approach

In this chapter, we treat FMEA as a group MCDM problem and obtain FMEA team members' opinions in the form of linguistic terms. Then, these linguistic terms are converted into triangular fuzzy numbers (cf. Tables 12.1 and 11.1). As a result, a

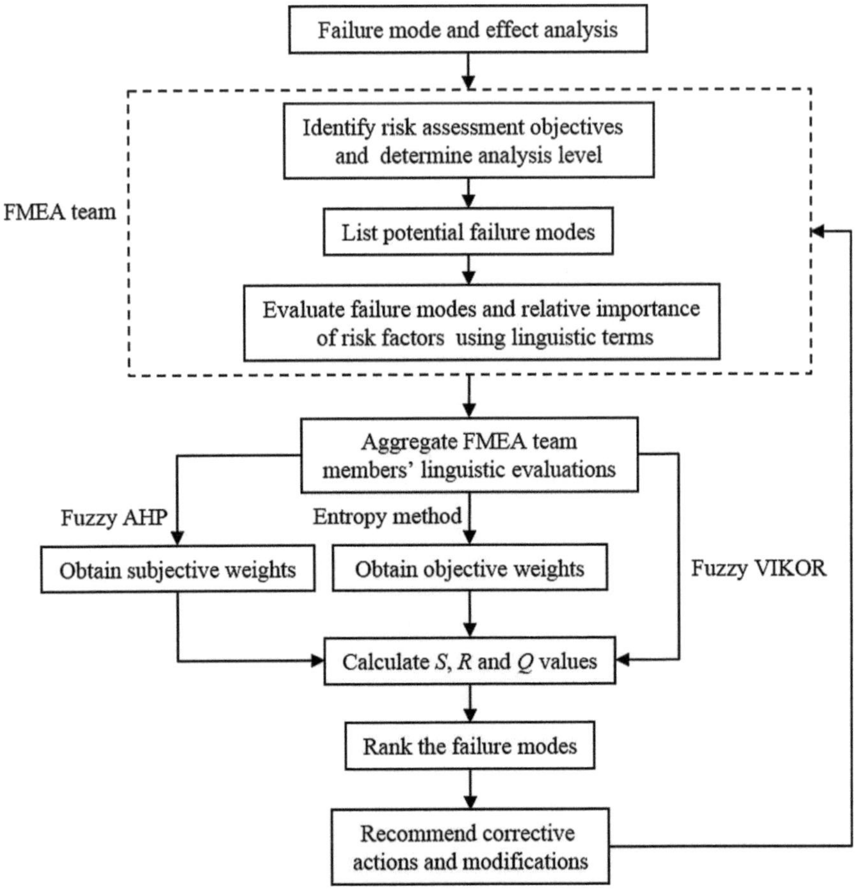

Fig. 12.2 Flowchart of the proposed FMEA approach (Liu et al. 2015)

systematic approach based on combination weighting and fuzzy VIKOR method is proposed to determine the risk priorities of failure modes in FMEA. All necessary steps required for making a fuzzy criticality assessment using the proposed approach are outlined in Fig. 12.2. These steps involved are explained in detail as follows (Liu et al. 2015):

Step 1. Identify the objectives of risk assessment process and determine the analysis level.

Step 2. Establish a FMEA team, list the potential failure modes, and describe a finite set of relevant risk factors.

Step 3. Determine appropriate linguistic terms for risk factors and their relative weights.

Step 4. Obtain the subjective weights of risk factors by using the fuzzy AHP approach:

- Each team member is asked to assign linguistic terms to the pairwise comparisons among risk factors.
- The team members' linguistic evaluations are aggregated to get fuzzy relative importance, and a fuzzy pairwise comparison matrix for risk factors is constructed.
- Consistency of the fuzzy pairwise comparison matrix is examined after defuzzification of the matrix according to the COA method.
- Fuzzy geometric means for risk factors are computed.
- Fuzzy weights of risk factors are calculated.
- The values of fuzzy weights are defuzzified and normalized to get the subjective weight of each risk factor.

Step 5. Obtain the objective weights of risk factors by using the entropy method:

- The team members' linguistic evaluations of each failure mode with respect to risk factors are defuzzified and normalized to get the projected outcomes.
- The entropy of the set of projected outcomes for each risk factor is computed.
- The divergence degrees of the intrinsic information for risk factors are defined.
- The subjective weights of all the risk factors are obtained.

Step 6. Calculate the S, R, and Q values by applying the fuzzy VIKOR approach:

- The team members' linguistic evaluations of failure modes with respect to each risk factor are aggregated.
- The fuzzy best f_j^* and the fuzzy worst f_j^- values are determined.
- Normalized fuzzy distances are calculated.
- The values S, R, and Q are calculated, respectively.

Step 7. Determine the risk priority orders of failure modes in terms of the values S, R, and Q in decreasing order.

Step 8. Analyze the results and take necessary corrective actions to improve the reliability and safety of the system.

12.3 An Illustrative Example

To demonstrate the proposed approach for the risk evaluation in FMEA, a real-world application in medical risk management is employed in this section.

12.3.1 Implementation

A tertiary care university hospital located in Shanghai, China, has applied the proposed FMEA as its technique to analyze the risk of general anesthesia process (Liu et al. 2015). The steps and analysis of this application example are given below. The hospital desires to identify the most serious failure modes during general anesthesia process to prevent the incidence of medical errors (Step 1). A FMEA team of five decision makers, DM1, DM2, DM3, DM4, and DM5, has been set up in the hospital in order to evaluate the failure modes in general anesthesia process. The decision makers included two anesthetists, two chief physicians, and one operating room nurse. Note that the five decision makers are assigned the following relative weights: 0.15, 0.20, 0.30, 0.25, and 0.10 to reflect their differences in performing the FEMA.

Seven potential failure modes have been identified by the FMEA team that included arterial gas bolt (FM1), visceral injury (FM2), respiratory depression (FM3), not check anesthesia equipment completely (FM4), not estimate surgery enough (FM5), blood transfusion wrong (FM7), and go esophageal (FM6). The risk factors, *O*, *S*, and *D*, were defined according to historical data and questionnaires answered by all team members (Step 2).

The five decision makers use the linguistic terms shown in Table 12.1 to assess the subjective importance of the risk factors. Also, they use the linguistic rating terms shown in Table 11.1 to evaluate the ratings of failure modes with respect to each risk factor. The evaluations of the five FMEA team members for the risk factors with respect to each failure mode are obtained as expressed in Table 12.2 (Step 3).

According to the fuzzy AHP method, the evaluations of FMEA team members in linguistic terms are used to calculate the subjective weights of risk factors by pairwise comparisons, and the results are given in Table 12.3 (Step 4). In this case study, the consistency ratio calculated is lower than 0.1 according to the experts' evaluations. Thus, the pairwise comparison matrix can be considered as consistent, and the survey is valid in terms of the fuzzy AHP.

Subsequently, the linguistic evaluations shown in Table 12.2 are converted into triangular fuzzy numbers. Then, the aggregated fuzzy ratings of failure modes are calculated to determine the fuzzy decision matrix, as in Table 12.4. Based on the entropy methodology, the objective weights of risk factors can be obtained, which are shown in Table 12.5 (Step 5). As noted in Table 12.5, the risk factor D has bigger weight than other risk factors, and in contrast, the O and S weights are very small.

In the next step, the fuzzy best f_j^* and the fuzzy worst f_j^- values of all risk factor ratings are determined by Eqs. (12.13)–(12.14) as follows:

$$\tilde{f}_O^* = (0.80, 2.60, 4.60), \quad \tilde{f}_S^* = (1.70, 3.70, 5.70), \quad \tilde{f}_D^* = (0.00, 0.25, 1.50),$$
$$\tilde{f}_O^- = (8.20, 9.40, 9.80), \quad \tilde{f}_S^- = (7.80, 9.40, 10.00), \quad \tilde{f}_D^- = (4.40, 6.40, 8.40).$$

Table 12.2 Ratings of failure modes assessed by the FMEA team members (Liu et al. 2015)

Failure modes	*O*					*S*					*D*				
	DM1	DM2	DM3	DM4	DM5	DM1	DM2	DM3	DM4	DM5	DM1	DM2	DM3	DM4	DM5
FM1	M	M	M	MH	M	ML	ML	ML	M	M	M	ML	ML	ML	ML
FM2	H	MH	H	MH	MH	H	MH	H	H	H	M	M	ML	M	M
FM3	VH	MH	VH	VH	VH	MH	MH	MH	MH	MH	MH	M	MH	MH	M
FM4	M	M	L	M	M	M	M	ML	M	M	VL	ML	VL	ML	VL
FM5	M	ML	M	M	M	M	MH	MH	M	M	L	ML	L	L	L
FM6	MH	H	M	MH	M	H	H	H	H	H	L	M	L	L	VL
FM7	ML	L	ML	ML	ML	VH	H	H	VH	H	VL	VL	VL	L	VL

Table 12.3 Subjective weights of risk factors by fuzzy AHP method (Liu et al. 2015)

	O	*S*	*D*	w^s
O	(1.000, 1.000, 1.000)	(0.480, 0.567, 1.700)	(0.840, 1.667, 3.400)	0.435
S	(0.588, 1.765, 2.083)	(1.000, 1.000, 1.000)	(1.500, 3.500, 5.500)	0.608
D	(0.294, 0.600, 1.190)	(0.182, 0.286, 0.667)	(1.000, 1.000, 1.000)	0.229

Table 12.4 Aggregated fuzzy ratings of failure modes (Liu et al. 2015)

Failure modes	*O*	*S*	*D*
FM1	(3.50, 5.50, 7.50)	(1.70, 3.70, 5.70)	(1.30, 3.30, 5.30)
FM2	(5.90, 7.90, 9.45)	(6.60, 8.60, 9.80)	(2.40, 4.40, 6.40)
FM3	(8.20, 9.40, 9.80)	(5.00, 7.00, 9.00)	(4.40, 6.40, 8.40)
FM4	(2.10, 3.80, 5.80)	(2.40, 4.40, 6.40)	(0.45, 1.35, 2.80)
FM5	(2.60, 4.60, 6.60)	(4.00, 6.00, 8.00)	(0.20, 1.40, 3.40)
FM6	(4.60, 6.60, 8.40)	(7.00, 9.00, 10.00)	(0.60, 1.70, 3.60)
FM7	(0.80, 2.60, 4.60)	(7.80, 9.40, 10.00)	(0.00, 0.25, 1.50)

Table 12.5 Objective weights of risk factors by entropy method (Liu et al. 2015)

	O	*S*	*D*
E	0.965	0.977	0.892
div	0.035	0.023	0.108
w^o	0.209	0.140	0.651

The normalized fuzzy distance can be calculated using Eq. (12.15) for each risk factor of the failure modes identified in the FMEA, as shown in Table 12.6. Then, the values of S, R, and Q are calculated for all failure modes as in Table 12.7 (Step 6). Finally, the risk priority orders of the failure modes by S, R, and Q in decreasing order are shown in Table 12.8 (Step 7).

From Table 12.8, it can be seen that the risk ranking of the seven failure modes is FM3 $\succ$ FM2 $\succ$ FM6 $\succ$ FM7 $\succ$ FM1 $\succ$ FM5 $\succ$ FM4 (Step 8). According to the comprehensive evaluation results, FM3 is the most serious failure mode and should be given the top risk priority by the hospital; this will be followed by FM2, FM6, FM7, FM1, FM5, and FM4.

Table 12.6 Normalized fuzzy distances of failure modes (Liu et al. 2015)

Failure modes	*O*	*S*	*D*
FM1	0.434	0.000	0.493
FM2	0.779	0.857	0.670
FM3	1.000	0.609	1.000
FM4	0.189	0.129	0.172
FM5	0.296	0.424	0.218
FM6	0.592	0.920	0.256
FM7	0.000	1.000	0.000

Table 12.7 Values of S, R, and Q for all failure modes (Liu et al. 2015)

	FM1	FM2	FM3	FM4	FM5	FM6	FM7
S	0.356	0.866	0.99	0.185	0.35	0.647	0.374
R	0.217	0.321	0.44	0.076	0.159	0.344	0.374
Q	0.3	0.759	1	0	0.216	0.656	0.527

Table 12.8 Ranking of failure modes by S, R, and Q in decreasing order (Liu et al. 2015)

	FM1	FM2	FM3	FM4	FM5	FM6	FM7
By S	5	2	1	7	6	3	4
By R	5	4	1	7	6	3	2
By Q	5	2	1	7	6	3	4

12.3.2 Sensitivity Analysis

In the proposed FMEA model, the parameter v has been introduced as weight of the strategy of the maximum group utility. It plays an important role in the risk prioritization of failure modes. Generally, the value of v is taken as 0.5. However, the parameter v can take any value from 0 to 1. Therefore, it is necessary to conduct a sensitivity analysis on v for validating the obtained results. The related results according to the value of v are illustrated in Fig. 12.3. As can be seen, the ranking orders of five failure modes are not influenced by the v value. This means that the risk priorities of these failures are the same in terms of both maximum group utility and minimum individual regret. This result shows that the obtained results of the proposed approach are robust and reliable. On the other hand, the ranking of FM2 is improved according to the increase of v value. This fact reveals that FM2 has higher

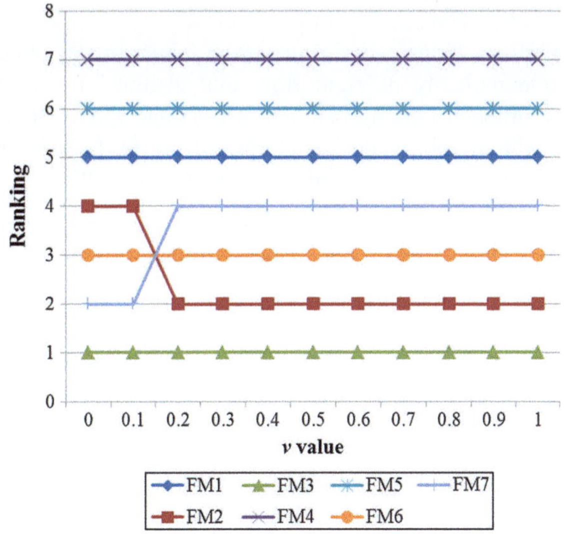

Fig. 12.3 Results of the sensitivity analysis (Liu et al. 2015)

level of risk when one focuses on maximum group utility. Also, the ranking of FM7 is high when the v value is small, indicating that its ranking is increased when the importance of minimum individual regret is increased. In other words, it is scored high-risk level when minimum individual regret is considered to be important.

12.3.3 Comparisons and Discussion

To further illustrate the effectiveness of the proposed fuzzy FMEA, the traditional FMEA model and the fuzzy TOPSIS approach (Kaya and Kahraman 2011) are considered. Also, to demonstrate the reflection of risk factor combination weights in the computing process, we perform the procedure of the proposed approach with considering only the subjective ($\varphi = 1$) or the objective weights ($\varphi = 0$) of risk factors on the application example. Table 12.9 exhibits the ranking comparison of the seven identified failure modes as obtained using these approaches.

Based on the information in Table 12.9, the findings can be summarized as follows (Liu et al. 2015):

- Failure modes can also be ranked when only the subjective or objective weights of risk factors are taken into account, but this may result in biased or even misleading ranking. For instance, FM6 turned out to be the second critical failure mode when considering only the subjective weights, while, considering only the objective weights, it ranks only at the fourth place, with a score of 0.224. At the same time, FM2 becomes the second critical one, with an overall score of 0.660.
- The final risk priority order can be definitely affected by the selection of the weight restriction φ. With the changing of the weight restriction from 1 to 0, the ranking orders of five of seven failure modes (71.4 %) are different, such as FM1, FM2, FM3, FM6, and FM7.
- The ranking order of the seven failure modes obtained by the fuzzy TOPSIS is remarkably different from that obtained by the proposed approach. A very different ranking is found in FM1 and FM7. The main reasons that brought the differences could be interpreted by the fact that the aggregation and normalization approaches of the VIKOR and TOPSIS methods are different.

On the other hand, the results of the proposed risk evaluation model and the conventional RPN method are somewhat similar. Except for FM1, FM5, and FM7, the risk priority ranks of the other failure modes provided by the proposed approach exactly match with those by the traditional FMEA. But the advantages of the proposed FMEA can be identified by a close look at the values of the risk factors for the failure modes with inconsistent rankings. For example, FM5 is ranked behind FM1 because it has a small detection rating in comparison with FM1. In the healthcare context, more weighting should be given to the risk factor of D due to the fact that healthcare failures may lead to serious injury or death to the patients once

Table 12.9 Ranking comparison (Liu et al. 2015)

Failure modes	Proposed approach ($\varphi = 1$)		($\varphi = 0.5$)		($\varphi = 0$)		Traditional FMEA					Fuzzy TOPSIS	
	Q	Ranking	Q	Ranking	Q	Ranking	O	S	D	RPN	Ranking	RC	Ranking
FM1	0.162	6	0.300	5	0.362	3	6	4	3	72	4	0.848	3
FM2	0.905	1	0.759	2	0.660	2	8	9	4	288	2	0.786	2
FM3	0.835	3	1.000	1	1.000	1	9	7	6	378	1	0.734	1
FM4	0.000	7	0.000	7	0.018	7	4	4	1	16	7	0.903	6
FM5	0.309	5	0.216	6	0.104	5	5	6	1	30	5	0.883	5
FM6	0.859	2	0.656	3	0.224	4	7	9	2	126	3	0.853	4
FM7	0.745	4	0.527	4	0.026	6	3	10	1	30	5	0.914	7

happened. Moreover, FM7 is successfully distinguished from FM5 in line with the proposed fuzzy FMEA. Both the two failure modes have the same detection rating, but the former has a very high severity rating and is therefore ranked higher than the latter. So, proposing FM7 as the fourth ranking and FM5 as the sixth ranking which is given by the proposed FMEA seems more genuine than those given by the traditional FMEA and the fuzzy TOPSIS methods.

The empirical example provided above has demonstrated that the proposed approach is an effective and useful tool to assess the risk of potential failure modes in fuzzy FMEA. In summary, compared with the traditional FMEA and its variants, the model proposed in this chapter has the following properties: (1) The proposed FMEA sufficiently considers different importance of risk factors. The risk factor weights are determined by combining the fuzzy AHP and the Shannon entropy measure, which makes the risk analysis result more consistent with the actual situation. (2) Based on the extended fuzzy VIKOR method, the proposed approach can simultaneously consider the maximum of group utility of the majority and the minimum of the individual regret for the opponent. The risk analysts may choose different coefficients of analysis mechanism (i.e., v) to rank failures according to their own subjective preferences. Thus, the proposed risk priority model using combination weighting and fuzzy VIKOR method is of efficiency and flexibility for FMEA.

References

Buckley JJ, Feuring T, Hayashi Y (2001) Fuzzy hierarchical analysis revisited. Eur J Oper Res 129(1):48–64

Chen CT (2000) Extensions of the TOPSIS for group decision-making under fuzzy environment. Fuzzy Sets Syst 114(1):1–9

Kaya T, Kahraman C (2011) Multicriteria decision making in energy planning using a modified fuzzy TOPSIS methodology. Expert Syst Appl 38(6):6577–6585

Liu HC, Liu L, Liu N, Mao LX (2012) Risk evaluation in failure mode and effects analysis with extended VIKOR method under fuzzy environment. Expert Syst Appl 39(17):12926–12934

Liu HC, You JX, You XY, Shan MM (2015) A novel approach for failure mode and effects analysis using combination weighting and fuzzy VIKOR method. Appl Soft Comput 28:579–588

Opricovic S (1998) Multi-Criteria Optimization of Civil Engineering Systems. Faculty of Civil Engineering, Belgrade

Saaty TL (1980) The Analytic Hierarchy Process. McGraw-Hill, New York

Shannon CE, Weaver W (1947) A Mathematical Theory of Communication. The University of Illinois Press, Urbana

Chapter 13
FMEA Combining VIKOR, DEMATEL, and AHP Methods

Liu et al. (2015a) developed a hybrid MCDM method for FMEA that combines VIKOR, DEMATEL (decision-making trial and evaluation laboratory), and AHP (analytic hierarchy process). A modified VIKOR method is employed to determine the effects of failure modes on together. Then, the DEMATEL technique is used to construct the influential relation map (IRM) among failure modes and causes of failures. Finally, the AHP approach based on DEMATEL is utilized to obtain the influential weights and give the prioritization levels for failure modes. The proposed FMEA can overcome the shortcomings and improve the effectiveness of the traditional FMEA. Particularly, it is able to capture the dependence and interactions between various failure modes and effects and provide guidance to analysts by setting the suitable maintenance strategies to improve the safety and reliability of complex systems.

13.1 The Proposed FMEA Model

To help risk analysts to formulate a more efficient and effective risk priority ranking, solving the problems concerning the traditional FMEA, Liu et al. (2015a) developed a new risk priority model using modified VIKOR method, DEMATEL technique, and AHP approach for the prioritization of the failure modes identified in FMEA. This model consists of three main stages: First, the modified VIKOR method is used to determine the effects of failure modes on together. Next, the DEMATEL technique is employed to construct the IRM among failure modes and causes of failures. Finally, the AHP approach based on DEMATEL is utilized to obtain the influential weights for the failure modes that have been individuated. By using the proposed FMEA, it is possible to determine how to improve failure modes and reduce the gaps to achieve the aspiration level and enhance the system

H.-C. Liu, *FMEA Using Uncertainty Theories and MCDM Methods*,
DOI 10.1007/978-981-10-1466-6_13

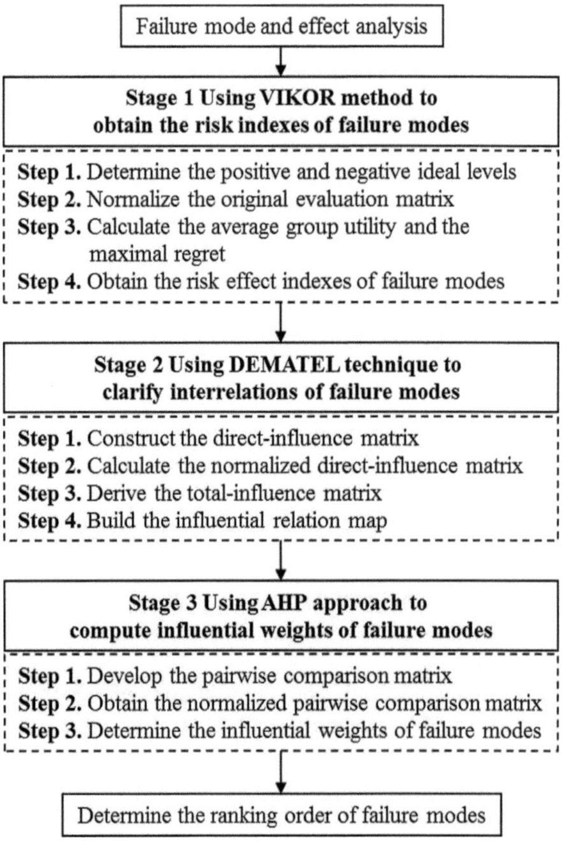

Fig. 13.1 Flowchart of the proposed FMEA model (Liu et al. 2015a)

reliability. The procedures of the hybrid MCDM approach are schematically shown in Fig. 13.1 and explained in detail in the following subsections.

13.1.1 The VIKOR Method for Determining Failure Effects

The VIKOR method was proposed by Opricovic (1998) as a MCDM method to solve the discrete decision problems with non-commensurable and conflicting criteria. It introduces a multi-criteria ranking index based on the particular measure of closeness to the ideal/aspired level solution (Liu et al. 2014). This ranking index is an aggregation of all criteria, the relative importance of criteria, and a balance between total and individual satisfaction. Using this concept, the VIKOR method can not only rank and select, but also improve alternatives for all criteria to achieve the aspired level.

In the first stage of the proposed FMEA model, the VIKOR is applied to determine the effects of failure modes on another and identify the gaps that each failure has to the aspired level. This method provides us with a road map to how we can improve upon each failure mode by minimizing the gap of each risk factor relative to the aspired level through remedial or corrective actions. Assuming that the potential failure modes identified in FMEA are represented by FM_1, FM_2, …, FM_m, the rating of the failure mode FM_i on the risk factor RF_j is denoted as f_{ij} (i = 1, 2, …, m; j = 1, 2, …, n); w_j is the weight of the jth risk factor, where j = 1, 2, …, n, and n is the number of risk factors. Then, the modified algorithm of VIKOR for FMEA is summarized as mentioned the following steps (Liu et al. 2015a):

Step 1. Determine the positive and negative-ideal levels

Suppose f_j^+, the positive-ideal level, represents the best value (aspired level) in each risk factor. In contrast, f_j^-, the negative-ideal level, represents the worst value in each risk factor. Equations (13.1)–(13.2) are then used to obtain the results.

$$f_j^+ = \text{the positive-ideal level,} \tag{13.1}$$

$$f_j^- = \text{the negative-ideal level.} \tag{13.2}$$

In FMEA, the evaluations of risk factors are obtained by using questionnaires with integer scales ranging from 1 to 10. Therefore, we can set the aspired level as $f_j^* = 0$ and the worst value as $f_j^- = 10$.

Step 2. Normalize the original evaluation matrix

The original risk evaluation matrix $F = \left[f_{ij}\right]_{m\times n}$ can be converted into a normalized gap-rating matrix $R = \left[r_{ij}\right]_{m\times n}$ (where the rating r_{ij} shows the gap of the failure mode FM_i regarding the risk factor RF_j) by

$$r_{ij} = \left(\left|f_j^* - f_{ij}\right|\right) / \left(\left|f_j^* - f_j^-\right|\right). \tag{13.3}$$

Step 3. Calculate the average group utility and the maximal regret

The average group utility S_i and the maximal regret R_i are computed through the following formulas:

$$S_i = \sum_{j=1}^{n} \left(w_j r_{ij}\right), \tag{13.4}$$

$$R_i = \max_j \left(w_j r_{ij}\right). \tag{13.5}$$

Step 4. Obtain the risk effect index

The risk effect index Q_i can be calculated using Eq. (13.6) for all the failure modes identified.

$$Q_i = v\frac{S_i - S^*}{S^- - S^*} + (1 - v)\frac{R_i - R^*}{R^- - R^*}, \tag{13.6}$$

where $S^* = \min_i S_i, S^- = \max_i S_i$, $R^* = \min_i R_i, R^- = \max_i R_i$ and v represents the weight for the strategy of maximum group utility, whereas $1 - v$ is the weight of the individual regret. When $S^* = 0$ and $R^* = 0$ (i.e., all risk factors have achieved the aspiration level) and $S^- = 1$ and $R^- = 1$ (i.e., the worst situation), Eq. (13.6) can be rewritten as

$$Q_i = vS_i + (1 - v)R_i, \tag{13.7}$$

where the range of v is $0 \leq v \leq 1$. Generally, the value of v is set to 0.5 (Liu et al. 2015a, b), which can be adjusted depending on the actual case considered. More specifically, $v = 1$ indicates that only the synthesized or integrated gap is considered, and $v = 0$ indicates that only the maximum weighted gap is used as the risk effect of failure mode.

13.1.2 The DEMATEL Technique for Building IRM

As stated in Sect. 11.1.2, the DEMATEL is a particularly pragmatic analytical method for visualizing the structure of complicated causal relationships. It is suitable to clarify the essential of a complex system and can help decision makers or risk analysts understand the relationships among potential failure modes. In this chapter, the DEMATEL is used to confirm the relationships between failure modes and causes of failures of a system to build the IRM among them (Liu et al. 2015a).

Step 1. Construct the direct-influence matrix

The direct-influence matrix Z is a $m \times m$ non-negative matrix obtained by pairwise comparisons in terms of influences and directions between elements in a system, in which z_{ij} represents the direct effect that element i has on element j.

$$Z = \begin{bmatrix} 0 & z_{12} & \cdots & z_{1m} \\ z_{21} & 0 & \cdots & z_{2m} \\ \vdots & \vdots & \vdots & \vdots \\ z_{m1} & z_{m2} & \cdots & 0 \end{bmatrix} \tag{13.8}$$

For the application of this matrix in FMEA, the direct-influence matrix Z can be constructed based on the relationships of failure modes and effects and the risk effect indexes determined by the VIKOR method.

Step 2. Calculate the normalized direct-influence matrix

Once the direct-influence matrix Z is developed, the normalized direct-influence matrix $X = \left[x_{ij}\right]_{m \times m}$ can be acquired by Eqs. (13.9) and (13.10).

$$X = \frac{Z}{s}, \tag{13.9}$$

where

$$s = \max\left\{ \max_{1 \le i \le m} \sum_{j=1}^{m} z_{ij}, \max_{1 \le j \le m} \sum_{i=1}^{m} z_{ij} \right\}. \tag{13.10}$$

All elements in matrix X are complying with $0 \le x_{ij} < 1$, $0 \le \sum_{i=1}^{m} x_{ij} < 1$, and $0 \le \sum_{j=1}^{m} x_{ij} < 1$, and at least one (but not all) row or column of the summation is equal to 1.

Step 3. Derive the total-influence matrix

Based on the normalized direct-relation matrix X, the total-influence matrix $T = \left[t_{ij}\right]_{m \times m}$ is determined via Eq. (13.11), in which I denotes the identity matrix.

$$T = X + X^2 + X^3 + \cdots + X^k = X(I - X)^{-1}, \quad \text{when}\, k \to \infty. \tag{13.11}$$

Step 4. Build the influential relation map

At this step, the sum of the rows and the sum of the columns from the total-influence matrix T are, respectively, expressed as the vectors R and C using Eqs. (13.12)–(13.13).

$$R = [r_i]_{m \times 1} = \left[\sum_{j=1}^{m} t_{ij}\right]_{m \times 1}, \tag{13.12}$$

$$C = [c_j]_{m \times 1} = \left[\sum_{i=1}^{m} t_{ij}\right]_{1 \times m}^{T}. \tag{13.13}$$

where r_i denotes the sum of the ith row of the matrix T and shows the sum of the direct and indirect effects that FM_i has on all the other failure modes. Similarly, c_j denotes the sum of the jth column of the matrix T and indicates the sum of direct and indirect effects that FM_j has received from all of the other failure modes.

Let $i = j$ and $i, j \in \{1, 2, \ldots, m\}$; the horizontal axis vector ($R + C$) is then defined by adding R to C, which illustrates the strength of influences that are given and received of the failure mode. Similarly, the vertical axis vector ($R - C$) is created by deducting C from R, which can divide the failure modes into a cause group and an effect group. In general, if ($R - C$) is positive, then FM_i has a net influence on the other failures and is part of the cause group; if ($R - C$) is negative, then FM_i is being influenced by the other failures on the whole and is part of the effect group. Therefore, an IRM can be achieved by mapping the data set of ($R + C$, $R - C$), which provides valuable information for the risk management decision making.

13.1.3 The AHP Approach for Ranking Failure Modes

The AHP, as presented by Saaty (1980), is a structured approach to solve MCDM problems by setting the priorities of alternatives. The main advantage of the AHP method is that it can create the chance of searching and evaluating the causal relationships between goals, factors, subfactors, and alternatives by breaking down the structure of the problem. Also, the use of AHP does not involve cumbersome mathematical calculations, which is easier to understand and can effectively manipulate both qualitative and quantitative criteria.

After the DEMATEL confirms the influential relationships between failure modes, the AHP approach is then applied to determine their influential weights and prioritize the failure modes accordingly (Liu et al. 2015a).

Step 1. Develop the pairwise comparison matrix

Based on the total-influence matrix T constructed by the DEMATEL technique, the pairwise comparison matrix T_C can be defined for the identified failure modes, as shown in Eq. (13.14).

$$T_C = \begin{bmatrix} t_C^{11} & t_C^{12} & \cdots & t_C^{1m} \\ t_C^{21} & t_C^{22} & \cdots & t_C^{2m} \\ \vdots & \vdots & \vdots & \vdots \\ t_C^{m1} & t_C^{m2} & \cdots & t_C^{mm} \end{bmatrix} \tag{13.14}$$

If 0 appears in the matrix, this means that the ith failure mode has no influence on the jth failure mode.

Step 2. Obtain the normalized pairwise comparison matrix

Then, the pairwise comparison matrix T_C is normalized by using the total degree of influence to obtain T_C^{α}, as shown by Eqs. (13.15) and (13.16).

$$d_j = \sum_{i=1}^{m} t_C^{ij}, \quad j = 1, 2, \ldots, m \tag{13.15}$$

$$\mathbf{T}_C^{\alpha} = \begin{bmatrix} t_C^{\alpha 11} & t_C^{\alpha 12} & \cdots & t_C^{\alpha 1m} \\ t_C^{\alpha 21} & t_C^{\alpha 21} & \cdots & t_C^{\alpha 2m} \\ \vdots & \vdots & \vdots & \vdots \\ t_C^{\alpha m1} & t_C^{\alpha m2} & \cdots & t_C^{\alpha mm} \end{bmatrix} \tag{13.16}$$

where $t_C^{\alpha ij} = t_C^{ij}/d_j$ represents the element of normalized influence for the element t_C^{ij} divided by the sum d_i^{11} of each column.

Step 3. Determine the influential weights of failure modes

The influential weight of the ith failure modes (w_i) can be derived by calculating the arithmetic mean of ith row and normalizing the arithmetic means of rows in the normalized pairwise comparison matrix T_C^{α}. That is,

$$t_C^{\alpha i} = \frac{1}{m}\sum_{j=1}^{m} t_C^{\alpha ij}, \tag{13.17}$$

$$w_i = t_C^{\alpha i} \Big/ \sum_{i=1}^{m} t_C^{\alpha i}. \tag{13.18}$$

Consequently, we can obtain the influential weights (i.e., global risk) for all the failure modes by using the AHP approach. For the FMEA problem, the higher the influential weight, the bigger the effect of the failure mode. Therefore, the failure modes identified in the FMEA can be prioritized or ranked according to the descending order of their influential weights w_i (i = 1,2, …, m).

13.2 An Illustrative Example

In this section, an empirical study of diesel engine's turbocharger system (Xu et al. 2002; Liu et al. 2015a) is used to illustrate the feasibility of the proposed FMEA model for the priority ranking of failure modes in the presence of interdependence among failure modes.

13.2.1 Implementation

The gas turbocharger of diesel engine utilizes the engine's exhaust gas pressure and heat energy to cause turbine wheel to rotate, which in turn causes the compressor wheel to compress the air–fuel mixture and deliver it under pressure to the combustion chamber of the engine. So, the denser charge in the combustion chamber can develop more horsepower during the combustion cycle. However, its operational conditions are very severe due to the high temperature (up to 700 °C) of exhaust gas and the high-speed rotation (up to 50,000 revolutions/min). Therefore, it is of critical importance to conduct a design FMEA of turbocharger to enhance its life and reliability. Via expert's knowledge and experience, all potential failure modes for the turbocharger and its components are shown in Table 13.1.

An FMEA team identifies eight main failure modes in operating the turbocharger and needs to prioritize them in terms of the risk factors O, S, and D so that high risky failures can be corrected with a top priority. Then, the FMEA team members are asked to determine the ratings of the eight failure modes on the three risk factors and the results are shown in Table 13.2. The relative weights of the risk factors are determined as $w_O = 0.35$, $w_S = 0.40$, and $w_D = 0.25$. Following the VIKOR algorithm described in the previous section, the normalized gap-rating matrix $R = \left[r_{ij}\right]_{8\times 3}$ and the risk effect index value Q_i for each failure mode are calculated by using Eqs. (13.3)–(13.6) as presented in Table 13.3.

Table 13.1 Failure modes for the turbocharger subsystem (Liu et al. 2015a)

Failure modes (system)	Components	Failure modes (components)	Effects on same level	Effects on next level
Damaged turbocharger (F_{01})	1. Turbine wheel	Blade heavy rubbing (F_{11})		Cause F_{03}, F_{04}
		Broken blade (F_{12})	Cause F_{34}, F_{41}, F_{42}, F_{43}	Cause F_{01}, F_{03}, F_{04}
		Deposited carbon on the blade (F_{13})	Cause F_{41}	Cause F_{03}, F_{04}
	2. Shaft	Worn (F_{21})	Cause F_{41}	Cause F_{01}, F_{04}
		Excessive deformation (F_{22})	Cause F_{43}	Cause F_{03}, F_{04}
		Broken (F_{23})	Cause F_{12}, F_{34}	Cause F_{01}, F_{03}, F_{04}
Oil leakage (F_{02})	3. Compressor wheel	Blade heavy rubbing (F_{31})		Cause F_{03}, F_{04}
		Nicked blade (F_{32})	Cause F_{41}	Cause F_{03}, F_{04}
		Deposited dirt on the blade (F_{33})	Cause F_{41}	Cause F_{03}, F_{04}
		Blade damaged (F_{34})	Cause F_{41}	Cause F_{01}, F_{03}, F_{04}
	4. Full-floating journal bearings	Worn bearing (F_{41})	Cause F_{11}, F_{12}, F_{101}, F_{111}	Cause F_{01}, F_{03}, F_{04}
		Broken bearing (F_{42})	Cause F_{101}, F_{111}	Cause F_{01}, F_{03}, F_{04}
		Bearing seizure (F_{43})	Cause F_{11}, F_{12}	Cause F_{01}, F_{03}, F_{04}
Loss of power output and excessive smoke (F_{03})	5. Thrust bearing and rings	Damaged (F_{51})	Cause F_{12}, F_{34}	Cause F_{01}
	6. Locknut	Fracture (F_{61})	Cause F_{12}, F_{34}	Cause F_{01}
	7. Bearing housing	Blocked oil inlet passage (F_{71})	Cause F_{41}, F_{42}, F_{43}	Cause F_{01}
		Blocked oil exit funnel (F_{72})	Cause F_{41}, F_{43}, F_{73}	Cause F_{02}, F_{03}
		Housing crack (F_{73})		Cause F_{01}
	8. Oil deflector	Damaged (F_{81})		Cause F_{02}

(continued)

Table 13.1 (continued)

Failure modes (system)	Components	Failure modes (components)	Effects on same level	Effects on next level
Noise (F_{04})	9. Heat shroud	Damaged (F_{91})	Cause F_{111}, F_{73}	Cause F_{01}
	10. Compressor sealing ring	Fracture (F_{101})		Cause F_{02}
		Leakage (F_{102})	Cause F_{13}	Cause F_{02}, F_{03}
	11. Turbine sealing ring	Fracture (F_{111})	Cause F_{41}, F_{43}	Cause F_{01}
		Leakage (F_{112})		Cause F_{03}, F_{04}
	12. Operator	Start and stop operation error (F_{121})	Cause F_{34}, F_{41}, F_{42}, F_{43}	Cause F_{01}, F_{03}, F_{04}

Table 13.2 FMEA for the turbocharger by RPN (Liu et al. 2015a)

Failure modes	*O*	*S*	*D*	RPN
F_{11}	2	5	3	30
F_{13}	2	4	2	16
F_{61}	2	8	8	128
F_{71}	5	7	2	70
F_{72}	2	6	2	24
F_{101}	2	7	4	56
F_{111}	2	6	3	36
F_{121}	3	7	8	168

Table 13.3 Normalized gap-rating matrix and risk effect index (Liu et al. 2015a)

Failure modes	*O*	*S*	*D*	Q_i
F_{11}	0.2	0.5	0.3	0.273
F_{13}	0.2	0.4	0.2	0.220
F_{61}	0.2	0.8	0.8	0.455
F_{71}	0.5	0.7	0.2	0.393
F_{72}	0.2	0.6	0.2	0.300
F_{101}	0.2	0.7	0.4	0.365
F_{111}	0.2	0.6	0.3	0.313
F_{121}	0.3	0.7	0.8	0.433

After the effects among failure modes are determined, the DEMATEL technique is adopted for construing the IRM. The direct-influence matrix Z in terms of influences and directions between the potential failure modes is established as shown in Table 13.4. Using the matrix Z, the normalized direct-influence matrix X is calculated through Eqs. (13.9) and (13.10). Then, Eq. (13.11) is used to derive

Table 13.4 The direct-influence matrix (Liu et al. 2015a)

	F_{11}	F_{12}	F_{13}	F_{34}	F_{41}	F_{42}	F_{43}	F_{73}	F_{101}	F_{111}	F_{1}	F_{2}	F_{3}	F_{4}
F_{11}													0.273	0.273
F_{12}				0.01	0.01	0.01	0.01				0.01		0.01	0.01
F_{13}					0.220								0.220	0.220
F_{21}					0.01						0.01			0.01
F_{22}							0.01						0.01	0.01
F_{23}		0.01		0.01							0.01		0.01	0.01
F_{31}													0.01	0.01
F_{32}					0.01								0.01	0.01
F_{33}					0.01								0.01	0.01
F_{34}					0.01						0.01		0.01	0.01
F_{41}	0.01	0.01							0.01	0.01	0.01		0.01	0.01
F_{42}									0.01	0.01	0.01		0.01	0.01
F_{43}	0.01	0.01									0.01		0.01	0.01
F_{51}		0.01		0.01							0.01			
F_{61}		0.455		0.455							0.455			
F_{71}					0.393	0.393	0.393				0.393			
F_{72}					0.300		0.300	0.300				0.300	0.300	
F_{73}											0.01			
F_{81}												0.01		
F_{91}								0.01		0.01	0.01			
F_{101}												0.365		
F_{102}			0.01									0.01	0.01	
F_{111}					0.313		0.313				0.313			
F_{112}													0.01	0.01
F_{121}				0.433	0.433	0.433	0.433				0.433		0.433	0.433

Table 13.5 Sum of influences given and received among the failure modes (Liu et al. 2015a)

Failure modes	R	C	$R + C$	$R - C$
F_{11}	0.180017	0.010076	0.190092	0.169941
F_{12}	0.023365	0.166971	0.190335	−0.143606
F_{13}	0.219835	0.003303	0.223138	0.216532
F_{21}	0.009992	0	0.009992	0.009992
F_{22}	0.009966	0	0.009966	0.009966
F_{23}	0.016636	0	0.016636	0.016636
F_{31}	0.006606	0	0.006606	0.006606
F_{32}	0.009992	0	0.009992	0.009992
F_{33}	0.009992	0	0.009992	0.009992
F_{34}	0.013296	0.303607	0.316902	−0.290311
F_{41}	0.025229	0.567101	0.592330	−0.541873
F_{42}	0.017951	0.276357	0.294307	−0.258406
F_{43}	0.017187	0.483282	0.500469	−0.466095
F_{51}	0.010030	0	0.010030	0.010030
F_{61}	0.456377	0	0.456377	0.456377
F_{71}	0.526406	0	0.526406	0.526406
F_{72}	0.499989	0	0.499989	0.499989
F_{73}	0.003303	0.102395	0.105698	−0.099092
F_{81}	0.003303	0	0.003303	0.003303
F_{91}	0.010957	0	0.010957	0.010957
F_{101}	0.120562	0.009392	0.129954	0.111169
F_{102}	0.010635	0	0.010635	0.010635
F_{111}	0.314040	0.012695	0.326735	0.301344
F_{112}	0.006606	0	0.006606	0.006606
F_{121}	1.010523	0	1.010523	1.010523
F_1	0	0.566627	0.566627	−0.566627
F_2	0	0.227392	0.227392	−0.227392
F_3	0	0.451344	0.451344	−0.451344
F_4	0	0.352253	0.352253	−0.352253

the total-influence matrix T and Eqs. (13.12) and (13.13) are utilized to find the sum of the influences given (R) and received (C) for every failure mode. The results so obtained are presented in Table 13.5. The influence relationships can be visualized by drawing an IRM of the eight component failure modes and the four system failure modes, as illustrated in Fig. 13.2.

As shown in Fig. 13.2, all the eight component failure modes have a positive ($R - C$) value which means they will affect other failures more than be affected by others. F_{121} has the largest positive value and thus has the greatest effect on the other failure modes. In addition, Table 13.5 shows that F_{121} has the highest ($R + C$) value. According to the R and C scores of F_{121}, its influential impact on others ranks first while the impact it receives from others is 0. It is indicated that

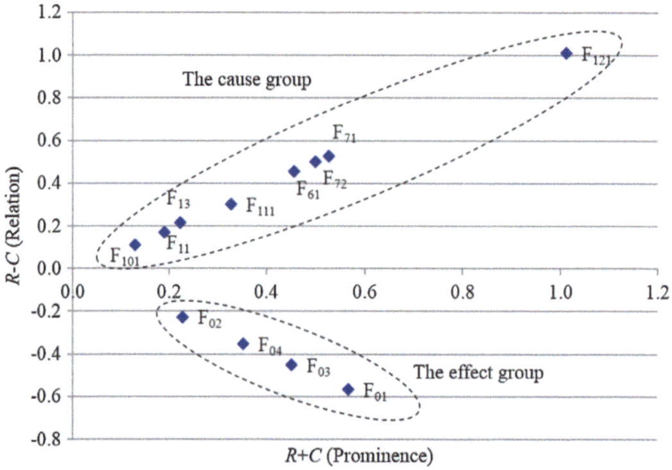

Fig. 13.2 Influential relation map among failure modes (Liu et al. 2015a)

F_{121} can dramatically affect the other failures, and that improvement of F_{121} can lead to the amelioration of the whole system. Therefore, F_{121} should be given the highest priority when initiating corrective actions to improve the reliability of the system. The $(R - C)$ value of F_{101} is positive, which suggests that F_{101} is a net cause failure mode for the whole system. But F_{101} has the lowest $(R + C)$ value in all the component failure modes, which implies that it is less significant than the other failures. These results show that F_{101} do not have enough power to improve the system and should be given a relative low priority in allocating maintenance resources.

In this chapter, the DEMATEL is combined with the AHP to obtain the influential weights of failure modes. This information is used to determine the risk priority order of the failure modes identified in FMEA. Based on the total-influence matrix *T* determined by the DEMATEL, the influential weight is calculated using the AHP method for each failure mode, as shown in Table 13.6. The DEMATEL-based AHP approach allows us to derive the global risk of the failure

Table 13.6 Influence weights and ranking of the eight major failure modes (Liu et al. 2015a)

Failure modes	Weights	Ranking
F_{11}	0.032496	8
F_{13}	0.040515	6
F_{61}	0.119630	3
F_{71}	0.103846	4
F_{72}	0.152725	2
F_{101}	0.037871	7
F_{111}	0.051294	5
F_{121}	0.199938	1

modes, which helps to understand the absolute risk of individual failures in an overall perspective. The eight failure modes are arranged according to their influential weights in descending order. The purpose is to determine the most important failure modes for preventive measures in order to help assure the safety and reliability. The results indicate that F_{121} is the first priority in terms of the global risk, which is followed by F_{72}, F_{61}, F_{71}, F_{111}, F_{13}, F_{101}, and F_{11}. Therefore, the complete risk ranking of the eight failure modes as obtained using the proposed model is given as: $F_{121} \succ F_{72} \succ F_{61} \succ F_{71} \succ F_{111} \succ F_{13} \succ F_{101} \succ F_{11}$.

Despite the yielded ranking of failure modes, the proposed hybrid MCDM model also helps to identify the risk gaps of the failure modes on each risk factor by using the modified VIKOR method. For example, the normalization of the risk factor scores of F_{121} in Table 13.3 demonstrated that the risk factor gap of *D* is 0.8 and the gap for the *S* risk factor is 0.7 constituting the largest gaps, which the risk analyst should improve as a priority. The normalization of the risk factor scores of F_{72} shows that the gap of the S risk factor is 0.6, constituting the largest gap, which the analyst should first improve. The normalization of the risk factor scores of F_{61} in Table 13.3 shows that the gaps of the *S* and *D* risk factors are 0.8 constituting the largest gap, which the analyst should improve as a priority. In the same manner, the risk analyst can refer to the normalized gap-rating matrix derived by the VIKOR method to prioritize the improvements of risk factors for the high risky failure modes and carry out suitable maintenance strategies to minimize recurrence of these failure modes.

13.2.2 Comparisons and Discussion

For further illustrating the effectiveness of the proposed model, the conventional RPN method, the fuzzy FMEA (Xu et al. 2002), and the DEMATEL-based FMEA (Seyed-Hosseini et al. 2006) have been implemented for the same case study. The ranking results of the eight failure modes as acquired using these approaches are shown in Table 13.7.

First, it can be seen that there are visible differences between the risk priority rankings produced by the four FMEA methods. This shows that the conventional RPN, the fuzzy FMEA, the DEMATEL-based FMEA, and the proposed model have different mechanisms in determining the risk priority ranking of failure modes. Both the conventional RPN and the fuzzy FMEA give the fourth rank to FM_{101}, which is more important than FM_{13}. However, by the proposed model and the DEMATEL-based method, FM_{13} has a higher risk in comparison with FM_{101}. This can be understood from the fact that the conventional RPN and the fuzzy FMEA methods did not consider the interrelationships among failure modes and effects, which had caused changes in the rankings of the failure modes. As analyzed in the former subsection and according to Tables 13.4 and 13.5, FM_{13} has a more influence than FM_{101} on the system and thus is more important. Also, the risk factor weights are ignored in both the conventional RPN and the fuzzy FMEA methods.

Table 13.7 Ranking comparisons (Liu et al. 2015a)

Failure modes	RPN	Fuzzy FMEA	DEMATEL	The proposed model
F_{11}	6	5	8	8
F_{13}	8	7	6	6
F_{61}	2	1	2	3
F_{71}	3	1	3	4
F_{72}	7	7	4	2
F_{101}	4	4	7	7
F_{111}	5	5	5	5
F_{121}	1	1	1	1

Moreover, in the fuzzy FMEA, the fuzzy if-then rules with the same consequent but different antecedents are unable to be distinguished from one another (Liu et al. 2013a, b, 2015a). As a result, there are many failure modes with the same rankings, which cannot be fully ranked and well distinguished from each other.

Second, the risk priority rankings of the failure modes F_{61}, F_{71}, and F_{72} yielded by the proposed model are different from those by the DEMATEL-based FMEA. These inconsistent ranking results are mainly because the conventional RPN was used to determine the severity of effect or influence between failure modes and effects in the DEMATEL-based method. Hence, it could not solve the shortcomings of the traditional FMEA. Furthermore, the DEMATEL-based FMEA only considered the net influence of the failure mode ($R - C$) in the risk prioritization process, which may cause biased conclusions. In contrast, the proposed model considers the type of relationships and severity of influences of failure modes all together and determines the risk ranking of failure modes by using the AHP method based on the results from the DEMATEL technique.

From the comparison with the listed methods, it can be concluded that the proposed model combining the VIKOR, the DEMATEL, and the AHP methods is an effective and efficient risk evaluation tool for the prioritization of failure modes in the system FMEA. Using the proposed hybrid MCDM approach, the relative weights of risk factors and the complex interactions and interdependences between the failure modes are incorporated in the failure analysis process. Moreover, the new proposed FMEA makes it possible for us to search for the failures' root causes (i.e., failure mechanisms) and set the suitable maintenance strategies to improve existing system reliability.

References

Liu HC, Liu L, Lin QL (2013a) Fuzzy failure mode and effects analysis using fuzzy evidential reasoning and belief rule-based methodology. IEEE Trans Reliab 62(1):23–36

Liu HC, Liu L, Liu N (2013b) Risk evaluation approaches in failure mode and effects analysis: A literature review. Expert Syst Appl 40(2):828–838

Liu HC, Ren ML, Wu J, Lin QL (2014) An interval 2-tuple linguistic MCDM method for robot evaluation and selection. Int J Prod Res 52(10):2867–2880

Liu HC, You JX, Ding XF, Su Q (2015a) Improving risk evaluation in FMEA with a hybrid multiple criteria decision making method. International Journal of Quality & Reliability Management 32(7):763–782

Liu HC, You JX, You XY, Shan MM (2015b) A novel approach for failure mode and effects analysis using combination weighting and fuzzy VIKOR method. Appl Soft Comput 28:579–588

Opricovic S (1998) Multi-Criteria Optimization of Civil Engineering Systems. Faculty of Civil Engineering, Belgrade

Saaty TL (1980) The Analytic Hierarchy Process. McGraw-Hill, New York

Seyed-Hosseini SM, Safaei N, Asgharpour MJ (2006) Reprioritization of failures in a system failure mode and effects analysis by decision making trial and evaluation laboratory technique. Reliability Engineering & System Safety 91(8):872–881

Xu K, Tang LC, Xie M, Ho SL, Zhu ML (2002) Fuzzy assessment of FMEA for engine systems. Reliability Engineering & System Safety 75(1):17–29

Appendix

Shortcomings of FMEA reported in the literature

Shortcomings	Frequency
The relative importance among *O*, *S*, and *D* is not taken into account. The three risk factors are treated with the same weight. This may not be the case when considering a practical application of FMEA For instance, a failure mode with a very high severity, low rate of occurrence, and moderate detectability (say 9, 3, and 5, respectively) may have a lower RPN (135) than one with all parameters moderate (say 5, 6, and 6 yielding an RPN of 180), even though it should have a higher priority for corrective action	72
Different combination of *O*, *S*, and *D* may produce exactly the same value of RPN, but their hidden risk implications may be totally different. This could entail a waste of resources and time, or in some cases, some high risky failure modes going unnoticed For example, two different failures with the values of 3, 4, 5 and 1, 10, 6 for *O*, *S*, and *D*, respectively, will have the same RPN value of 60. However, the hidden risk implications of the two failures may be very different because of the different severities of the failure consequence Consider two different failures having values of 3, 5, 2 and 2, 3, 5 for *O*, *S*, and *D*, respectively. Both these failures will have a total RPN of 12, however, their risk implications may not necessarily be the same	51
The three risk factors are difficult for FMEA team members to precisely determine. Much information in FMEA is often uncertain or vague and can be expressed in a linguistic way such as *likely*, *important*, or *very high* Usually, an assessment of the three risk factors is subjective and qualitatively described as natural language It is the difficulty or even impossible to precisely determine the probability of occurrence of failure modes It is usually difficult and inaccurate to give a "direct" and correct numerical evaluation of intangible quantities such as *O*, *S*, and *D* In particular, a highly technical system usually has no crisp inputs or outputs, and the relationships between the associated failure modes and effects are very complex, both subjectively and qualitatively It does not take into account linguistic variables that come from experts, which are in the FMEA team	40

(continued)

H.-C. Liu, *FMEA Using Uncertainty Theories and MCDM Methods*,
DOI 10.1007/978-981-10-1466-6

(continued)

Shortcomings	Frequency
The mathematical formula for calculating RPN is debatable and lacks a complete scientific basis. There is no rationale as to why *O*, *S*, and *D* should be multiplied to produce the RPN value	26
The direct and indirect relationships between potential failure modes and causes of failures are not taken into consideration Interdependencies among various failure modes and effects on the same level and different levels of hierarchical structure of an engineering system are not taken into account The conventional RPN methodology has not considered indirect relations between components and is deficient for systems with many subsystems or components When one cause of failure makes several potential failure modes, this cause should have a higher priority for a corrective action than others Traditionally, FMEA only considers the impact of single failure on the system. For large and complex systems, since multiple failures of components exist, assessing multiple failure modes with all possible combinations is impractical It is not likely to combine multiple qualitative assessments and is even more difficult to obtain the probability distributions that several failure modes occur simultaneously using traditional approach	23
The three risk factors *O*, *S*, and *D* are evaluated according to discrete ordinal scales of measure. But that the calculation of multiplication is meaningless on ordinal scales. Thus, the results obtained are not only meaningless but in fact misleading The original ordinal scale is transformed in a new cardinal scale characterized by a metric and by the integer number composition properties The assessment of RPN does not satisfy the usual requirements of measurement The evaluation of RPN is different from traditional concepts of quality measurement	22
The RPN considers only three risk factors mainly in terms of safety. Other important risk factors such as economical aspects are ignored The traditional FMEA only considers the *O*, *S*, and *D* risk factors, but other factors, such as failure cost, might have to be included to approximate the actual situation The RPN neglects production cost, quality, and/or other economical issues The traditional RPN ignores the effect of production quantity	21
The conversion of scores is different for the three risk factors. The relationship between *O* and the associated ratings is nonlinear, while the *D* and the associated ratings have a linear relationship For example, if the fault score is 3 and the detection score is 4, we get a 12. If the fault score is 4 and the detection score is 3, we also get 12, but in the first case the probability of a customer getting a faulty part is around 0.00003, while in the second it is about 0.0001	17
RPNs are not continuous with many holes. Many of the numbers in the range of 1–1, 000 cannot be formed from the product of *O*, *S*, and *D* and only 120 of the 1000 numbers can be generated	14

(continued)

(continued)

Shortcomings	Frequency
The RPN elements have many duplicate numbers. The total combinations of 1000 RPNs contain items that are not unique, some being repeated many times Most of the unique numbers can be formed in several different ways (only six RPN values are formed by a single, unique combination of *O*, *S*, and *D*) Nearly every RPN value is non-unique, some being recycled as many as 24 times. For example, 60 can be formed from 24 different combinations of *O*, *S*, and *D*	14
The mathematical form adopted for calculating RPN is strongly sensitive to variations in risk factor evaluations. Small variations in one rating may lead to vastly different effects on the RPN, depending on the values of the other risk factors For example, if *O* and *D* are both 10, then a 1-point difference in severity rating results in a 100-point difference in the RPN; if *O* and *D* are equal to 1, then the same 1-point difference results in only a 1-point difference in the RPN; if *O* and *D* are both 4, then a 1-point difference produces a 16-point difference in the RPN	14
The RPN method is only measuring from the risk viewpoint while ignoring the importance of corrective actions. The RPN cannot be used to measure the effectiveness of corrective actions The RPN cannot be used to analyze and evaluate the economical consequences of a failure with respect to the incurred costs to improve the final reliability level of the product The variable of the corrective action cost is not considered in the analysis The corrective actions may be interdependent; hence, if the implementation of corrective actions is in proper order, selection may maximize the improvement effect, bring favorable results in the shortest time, and provide the lowest cost	12
The RPN scale itself has some non-intuitive statistical properties. The initial and correctly assumed observation that the scale starts at 1 and ends at 1000 often leads to incorrect assumptions regarding the middle of the scale The definition of RPN on a formally wider scale than that of the three component risk factors, which generates a fictitious increase of its resolution	11
RPNs are heavily distributed at the bottom of the scale from 1 to 1000. This causes problems in interpreting the meaning of the differences between different RPNs For example, is the difference between the neighboring RPNs of 1 and 2 the same or less than the difference between 900 and 1000?	8
The customers' points of view are not considered during the risk analysis. Severity rates are determined only with respect to organization's point of view, not according to its customers The customers' expectations are not considered and only the manufacturer's point of view is taken into account when the prioritization of reliability efforts is actually decided	3
The cost of failure is not considered explicitly The three risk factors used for RPN calculation are ordinal scale variables. The cost due to failure cannot be defined by these ordinal numbers The evaluation of failure modes within the FMEA is insufficient, because the results do not reflect the cost that arises from a failure	3

(continued)

(continued)

Shortcomings	Frequency
The effort and resource outlay needed to implement FMEA is considered to be very substantial The realization and maintenance of a FMEA involves huge effort (long brainstorming sessions, costs of staff training, etc.) FMEA is a labor-intensive, time-consuming, and very tedious activity; hence, it is highly prone to errors	3
The attribution of values to *O*, *S*, and *D* is entirely arbitrary and can lead to incorrect final considerations, due more to the subjectivity of the analyst than the failure's objective characteristics Measuring severity and detection difficulty is very subjective and with no universal scale For example, two FMEAs compiled by different analysts, often result in varying RPNs for the same failure mode	3
The assumption that the scales of the *O*, *S*, and *D* risk factors have the same metric and that the same danger level corresponds to the same values on different index scales	3
The conventional RPN method does not consider the ordered weight	3
The diversity and ability of the team are the most important considerations, followed by training for the team member. This leads to a high cost. Furthermore, industrial practitioners usually find it hard to share their experience among team members of different background	2
The FMEA scales for *S* and *D* are basically only qualitative For example, a rank 8 severity may not be twice as severe as a rank 4 severity; the ratings are treated as if they represent numeric quantities	2
The result of the conventional RPN method is that it will lose some information provided by experts, which may cause biased conclusions For example, suppose that there are four experts to point out the *S* of the two failure modes. Failure mode 1 has an *S* value of 5 (each expert pointed out value are 5, 5, 4, and 4, respectively), and failure mode 2 has an *S* value of 5 (each expert pointed out value are 5, 6, 5, and 5, respectively); thus, they have the same *S* value of 5 for the two failure modes. However, in practice, failure mode 2 is more serious than failure mode 1	2
The risk factors and their interrelations are not considered in the failure modes evaluation process	1
The relative importance of experts cannot be included in the classical RPN calculation	1
Issues with incomplete information cannot be fully addressed using the traditional RPN methodology	1
The conventional RPN method does not consider the situation parameter, which is reflect a decision maker's current degree of optimism	1
As for the *S* parameter, in terms of graduation, there is a problem in determining the ranking which is more similar to an arrangement than a precise classification. Indeed, a value of 3 indicates a worse situation than that of 2, which can also be said for the values 7 and 8, and yet the two situations are not comparable and therefore judgment can overlook meaning	1

(continued)

(continued)

Shortcomings	Frequency
The numeric data interpretation brings about the simplification of the RPN calculation; however, it also increases the risk of moving its meaning away from the logic of the design team that supplied the figures	1
Various FMEA members score the impact of a risk factor on a failure mode into different criticality levels and finally take the average of these values. Hence, the criticality value of one risk factor may be same with other	1
The RPN calculation erroneously oversimplified that a twofold increase in a risk factor (e.g., severity) can be offset by a corresponding decrease of half in another risk factor	1

Frequency Number of articles which pointed out the shortcoming of the traditional FMEA in the literature